Genetic Modification

ISSUES

Volume 87

Editor

Craig Donnellan

First published by Independence
PO Box 295
Cambridge CB1 3XP
England

British Library Cataloguing in Publication Data
Genetic Modification – (Issues Series)
I. Donnellan, Craig II. Series
363.1'92

ISBN 1 86168 288 3

Printed in Great Britain
MWL Print Group Ltd

Typeset by
Claire Boyd

Cover
The illustration on the front cover is by Pumpkin House.

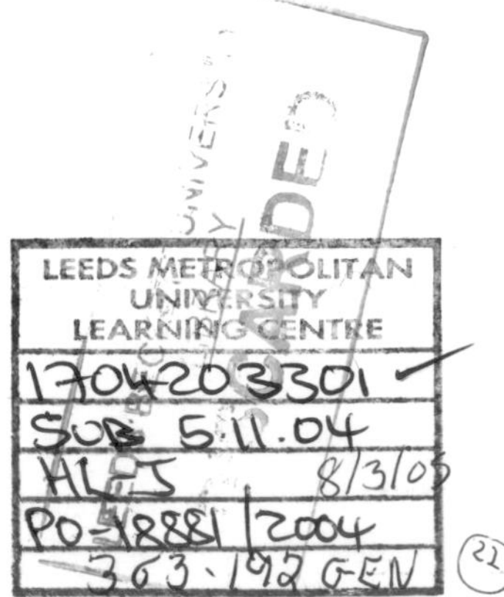

CONTENTS

Overview

Chapter One: GM Food

Introduction

Genetic Modification is the eighty-seventh volume in the ***Issues*** series. The aim of this series is to offer up-to-date information about important issues in our world.

Genetic Modification examines the debate about genetically modified food and organisms.

The information comes from a wide variety of sources and includes:
Government reports and statistics
Newspaper reports and features
Magazine articles and surveys
Website material
Literature from lobby groups
and charitable organisations.

It is hoped that, as you read about the many aspects of the issues explored in this book, you will critically evaluate the information presented. It is important that you decide whether you are being presented with facts or opinions. Does the writer give a biased or an unbiased report? If an opinion is being expressed, do you agree with the writer?

Genetic Modification offers a useful starting-point for those who need convenient access to information about the many issues involved. However, it is only a starting-point. At the back of the book is a list of organisations which you may want to contact for further information.

*****

OVERVIEW

GM basics

Information from the Food Standards Agency

What is GM?

GM, which stands for genetic modification or genetically modified, is the technique of changing or inserting genes. Genes carry the instructions for all the characteristics that an organism – a living thing – inherits. They are made up of DNA. Genetic modification is done either by altering DNA or by introducing genetic material from one organism into another organism, which can be either a different variety of the same or a different species. For example, genes can be introduced from one plant to another plant, from a plant to an animal, or from an animal to a plant. Transferring genes between plants and animals is a particular area of debate.

Sometimes the term 'bio-technology' is used to describe genetic modification. This also has a wider meaning of using micro-organisms or biological techniques to process waste or produce useful compounds, such as vaccines.

Why is genetic modification being used?

Genetic modification allows plants, animals and micro-organisms, such as bacteria, to be produced with specific qualities more accurately and efficiently than through traditional methods. It also allows genes to be transferred from one species to another to develop characteristics that would be very difficult or impossible to achieve through traditional breeding.

People have been breeding animals and new varieties of plants for many hundreds of years to develop or avoid certain qualities. Examples include racehorses that are bred to be faster and stronger, and roses, bred to give us a wider range of colours and to make them more resistant to disease. Over many generations, sometimes for thousands of years, the world's main food crops have been selected, crossed and bred to suit the conditions they are grown in and to make them tastier.

For example, cattle are bred according to whether they are for beef or dairy herds. Most of today's dairy cattle are very different from the cattle that were originally domesticated. Over the years, dairy herd breeding has focused on increasing milk yield and on improving the quality of the milk.

But whereas traditional methods involve mixing thousands of genes, genetic modification allows just one individual gene, or a small number of genes, to be inserted into a plant, or animal, to change it in a pre-determined way. Through genetic modification, genes can also be 'switched' on or off to change the way a plant or animal develops.

For example, herbicides are used to kill weeds in fields of crops but they can also affect the growth of the crops they are intended to protect. By using genetic modification, a gene with a particular characteristic, such as resistance to a specific herbicide, can be introduced into a crop plant. When that herbicide is sprayed on the field to kill the weeds, it will not hinder the growth of the crops.

Similarly, genetic modification can be used to reduce the amount of pesticide needed by altering a plant's DNA so it can resist the particular insect pests that attack it. Genetic modification can also be used to give crops immunity to plant viruses or to improve the nutritional value of a plant. In animals intended for food, genetic modification could potentially increase how fast and to what size they grow.

What is DNA?

DNA stands for deoxyribonucleic acid. It is the genetic material contained in the cells of all living things and it carries the information that allows organisms to function, repair and reproduce themselves.

Every cell of plants, micro-organisms (such as bacteria), animals, and people contains many thousands of different genes, which are made of DNA. These genes determine the

characteristics, or genetic make-up, of every living thing, including the food we eat. When we eat any food, we are eating the genes and breaking down the DNA present in the food.

DNA is made up of two separate strands of what are called 'nucleotides'. These are the building blocks of DNA and are twisted around each other in a double helix structure (see illustration below). The identity of a gene and the function it performs are determined by the number of nucleotides and the particular order in which they are strung together on chromosomes – this is known as the 'sequence' of the gene. Chromosomes are the cell structures that carry the DNA.

How does genetic modification work?

Genetic modification involves inserting or changing an organism's genes to produce a desired characteristic.

Inserting genes

When a plant, for example, is modified by inserting a gene from another plant into it, this is the process:

1. A plant that has the desired characteristic is identified.
2. The specific gene that produces this characteristic is located and cut out of the plant's DNA.
3. To get the gene into the cells of the plant being modified, the gene needs to be attached to a carrier. A piece of bacterial DNA called a plasmid is joined to the gene to act as the carrier.
4. A type of switch, called a 'promoter', is also included with the combined gene and carrier. This helps make sure the gene works properly when it is put into the plant being modified. Only a small number of cells in the plant being modified will actually take up the new gene. To find out which ones have done so, the carrier package often also includes a marker gene to identify them.
5. The gene package is then inserted back into the bacterium, which is allowed to reproduce to create many copies of the gene package.

DNA

The double helix structure of DNA, showing the base pairs

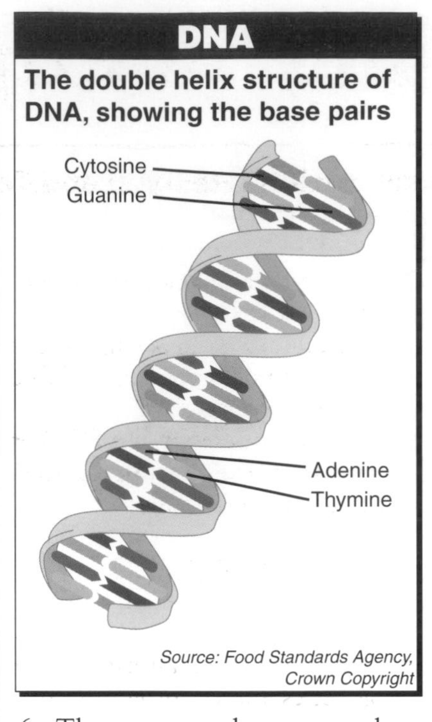

Source: Food Standards Agency, Crown Copyright

6. The gene packages are then transferred into the plant being modified. This is usually done in one of two ways:
   – by attaching the gene packages to tiny particles of gold or tungsten and firing them at high speed into the plant tissue. Gold or tungsten are used because they are chemically inert – in other words, they won't react with their surroundings
   – by using a soil bacterium, called *Agrobacterium tumefaciens*, to take it in when it infects the plant tissue. The gene packages are put into *A. tumefaciens*, which is modified to make sure it doesn't become active when it is taken into the new plant.
7. The plant tissue that has taken up the genes is then grown into full-size GM plants.
8. The GM plants are checked extensively to make sure that the new genes are in them and working as they should. This is done by growing the whole plants, allowing them to turn to seed, planting the seeds and growing the plant again, while monitoring the gene that has been inserted. This is repeated several times.

Altering genes

Genetic modification does not always involve moving a gene from one organism to another. Sometimes it means changing how a gene works by 'switching it off' to stop something happening. For example, the gene for softening a fruit could be switched off so that although the fruit ripens in the normal way, it will not soften as quickly. This can be useful because it means that damage is minimised during packing and transportation.

Controlling this gene 'switch' may also allow researchers to switch on modified genes in particular parts of a plant, such as the leaves or roots. For example, the genes that give a plant resistance to a pest might only be switched on in the bit of the plant that comes under attack, and not in the part used for food.

■ The above information is from the Food Standards Agency's website which can be found at www.foodstandards.gov.uk

Plant cell

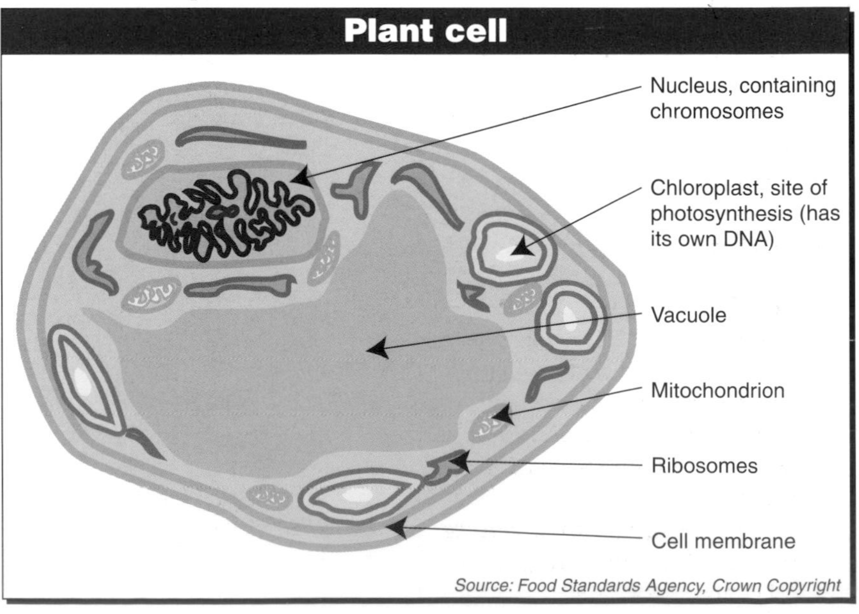

Source: Food Standards Agency, Crown Copyright

Genetically modified food

Information from the Society, Religion and Technology Project

Some see genetically modified crops opening up opportunities in agriculture, food and medicine. For some it's a threat to something basic about ourselves and the natural world – harmful, unnecessary, and benefiting big business at others' expense. But for many people, there are so many competing claims it's hard to know what to think. So what are the main issues?

The objections

Should we be modifying genes at all?

- It's 'playing God' or unnatural to mix genes from radically different organisms.
- We don't know enough about what we're doing to switch genes around in food.

GM is too risky

- We have not done enough to evaluate the risks to health or the environment.
- GM varieties can reduce biodiversity.
- GM genes will spread to non-GM varieties.
- GM contamination threatens organic farmers.

We don't need genetically modified food

- Agriculture is already too technological. This will only make it worse.
- There are better ways to improve pest resistance and reduce chemicals on the land.
- GM is no advantage to farmers if consumers don't want GM food.

It's bad for developing countries

- GM is just going to provide luxuries for rich, and won't feed the poor and hungry.
- Poor farmers need sustainable incomes not multi-national high tech seeds.
- GM is a technical fix diverting resources from exploring better indigenous solutions.

It's bad for democracy

- Most people don't want GM crops grown.
- Big business is imposing on our freedom under the guise of free trade.
- We must not allow the WTO or EC to force unwanted GM products on UK citizens.

The case in favour

We shouldn't be afraid of biotechnology

- We have been altering plant genetics for centuries through selective breeding.
- Adding 2 genes doesn't violate an organism.
- Why draw the line at GM, not elsewhere?

Risks are no worse than existing foods

- We have many safeguards in place.
- Health risks have been badly exaggerated.
- GM can improve the environment by using less chemicals and less soil tillage.
- With due care, GM and organic can co-exist.

Look at the opportunities for good

- Better resistance to weeds, pests, disease, yields, more efficient use of land.
- Better texture, flavour, nutritional value, longer shelf life, easier shipment.
- GM can offer farmers reduced costs to compete on a global market.

GM crops can help developing countries

- GM is a tool we will need to help feed a growing world population.
- China and India are already growing GM on a large scale.
- 'Golden rice' offers a potential solution to vitamin A deficiency in poor communities.

The democratic case

- The public don't rule out future uses of GM.
- With labelling, adequate protection can be given for those who object.
- Britain can't object to GM at the EC or WTO without clear scientific evidence.

Should we be doing genetic modification?

Some Christians object in principle to genetically modified food, as an unacceptable intervention in God's creation, violating barriers in the natural world. Others think using God's gift of our technical skills to change one or two genes is not wrong in itself, unless the change caused a major disruption in the organism. Such basic changes in genes and food require due precaution on food safety and environmental risk, but not out of proportion. SRT has an information sheet on GM animal issues.

The GM crisis

A UK GM tomato paste, introduced in 1996 and labelled in response to consumer advice, sold well until public attitudes changed in 1999. It began when the EU accepted US

GM soya and maize imports used to make soya oil and maize flour for many processed foods. The companies refused to segregate supplies or label products as GM, being more concerned with winning markets than public attitudes. Questions were also raised about potential risks to health, gene flow to non-GM crops, and a loss of biodiversity. When people realised they were eating GM foodstuffs whether they liked it or not, with perceived risks but no tangible benefits, and with no say in the decisions, a consumer backlash wasn't surprising. This led to an unofficial UK moratorium on growing GM crops till 2004.

Taking the public seriously

The mistake was in failing to see that making novel genetic changes in what we are offered to eat requires great care to listen to the public and to respect their views. The 2003 'GM Nation?' debate and other surveys show there is no public mandate to begin growing GM crops commercially in the UK yet. Most people seem not to be fundamentally opposed to GM as such, so much as sceptical and want more reliable evidence about long-term risks. Many think that future GM applications might offer benefits in medicine, developing countries or economically, but are suspicious about the role of big business and 'free' trade. Most GM so far has been for production efficiency to benefit seed companies and farmers not consumers. Those with basic objections to GM food must be given the option of not eating it, and should not have to pay more for what till now has been 'normal' food. EU legislation now requires labelling for all foods where GM processes have been used. And if there is no current public mandate we should not have GM products forced on us by the EC or the WTO.

Re-evaluating GM crops – government studies and trials in 2003

Government reports in 2003 concluded that economic benefits of GM crops to UK farmers would be depend on consumer acceptance, and that scientific risks should be weighed up, case by case. Thus the Government rejected the growing of herbicide tolerant GM oil seed rape and sugar beet after the farm scale trials showed these reduced field biodiversity compared with non-GM crops. It allowed a company a licence for GM maize for animal feed, which improved biodiversity, albeit compared with an old, aggressive weed killer. In doing so the Government ignored the results of its own public consultation, undermining its purpose. It considers that scientific evidence, not public opinion, is the only valid basis to reject a crop at the EC. But, for now, no GM will be grown commercially in the UK. The company abandoned its plans because of uncertainties about liability and its economic case.

Will genetic engineering really 'feed the world'?

Many Christians are concerned that the driving forces of biotechnology create products for western indulgence, neglecting real food shortages elsewhere in the world. The causes of hunger are more about poverty, war, political and social issues than inefficient production. Often better answers may come from better breeding with indigenous resources, than high tech solutions. Yet GM might help in some situations. GM vitamin A rice might help malnourished communities with no access to fresh vegetables. If genes could be altered to enable staple crops to grow in marginal conditions, it might make a difference to countries which struggle to feed themselves. But useful applications are often hard to engineer and offer no profits to private industry. GM has so far mostly been rich man's technology. To be serious about 'feeding the world' means radically reorienting research investment to put top priority on meeting the specific needs of marginal agriculture, using a diversity of old and new technologies. GM might be one tool amongst many. But will anyone take up that challenge?

■ For information sheets on other issues, contact the Society, Religion and Technology Project. See page 41 for their address details.

GM crops

Commercial cultivation of GM crops worldwide in 2003 (in millions of ha)[1]

Country	1998	1999	2000	2001	2002	2003
USA	20.5	28.7	30.3	35.7	39.0	42.8
Argentina	4.3	6.7	10.0	11.8	13.5	13.9
Canada	2.8	4.0	3.0	3.2	3.5	4.4
Brazil	0.0	0.0	0.0	0.0	0.0	3.0*
China	<0.1	0.3	0.50	1.5	2.1	2.8
Australia	0.1	0.1	0.15	0.21	0.1	0.1
South Africa	<0.1	0.1	0.20	0.27	0.3	0.4
Mexico	<0.1	<0.1	<0.1	<0.1	<0.1	<0.05
Spain	<0.1	<0.1	<0.1	<0.1	<0.1	<0.05
France	<0.1	<0.1	<0.1	0.0	0.0	0.0
Germany	0.0	<0.1	<0.1	<0.1	<0.1	<0.05
Bulgaria	0.0	0.0	0.0	0.0	0.0	<0.05
Colombia	0.0	0.0	0.0	0.0	<0.05	<0.05
Honduras	0.0	0.0	0.0	0.0	0.0	<0.05
India	0.0	0.0	0.0	0.0	<0.1	0.1
Indonesia	0.0	0.0	0.0	0.0	0.0	<0.05
Philippines	0.0	0.0	0.0	0.0	0.0	<0.05
Portugal	0.0	<0.1	<0.1	0.0	0.0	0.0
Romania	0.0	<0.1	<0.1	<0.1	<0.1	<0.05
Uruguay	0.0	0.0	<0.1	<0.1	<0.1	<0.05
Ukraine	0.0	<0.1	0.0	0.0	0.0	0.0
Total	**27.8**	**39.9**	**44.2**	**52.6**	**58.7**	**67.7**

1 James (2003) Preview: *Global status of commercialised transgenic crops: 2003.* ISAAA Briefs No. 30. ISAAA: Ithaca, NY.
* Estimated figure

Source: GeneWatch UK

GM dilemmas

Consumers and genetically modified foods

'Consumers will want reassurances on a range of issues, from product safety to environmental or ethical concerns. A drawing up of controls and codes of practice in this field should reflect those demands. More than ever before consumers will want to know exactly what they are eating and why.'

Quote from 1989 *Which? Way to Health* magazine when Consumers' Association (CA) first reported on genetic modification (GM).

Little has changed since then – consumers still want to know what they are eating and why. And there is still a real need for information as GM crops are being grown on a large scale across the world with many more products in the pipeline. Legislation requiring product approval and labelling of some ingredients is in place in the EU but it does not yet go far enough to address consumer concerns.

These concerns must at last be dealt with. The government has finally announced that there will be a public debate on issues around GM. Properly conducted it should provide ample opportunity for the government to understand consumer attitudes, address outstanding consumer concerns and assess the likely reaction to any potential products, including those that may offer consumer benefits and those that involve modifications to animals, fish and micro-organisms as well as plants. The debate offers all stakeholders the opportunity to stand back and consider what contribution, if any, GM can make to UK food production and what conditions would need to be met before consumers would accept further GM products. It must therefore explore consumer attitudes to future applications of the technology and the limits of consumer acceptability. It is also essential that the debate helps to determine the government's decision about whether or not to go ahead with commercial growing of GM crops. However, with the limited resources put aside for the debate, and a lack of commitment from the government to take the findings into account, we question whether this is a genuine debate. In many ways it appears that the government has already made its mind up that consumers have to accept GM foods.

Consumer concerns

Consumer concerns about GM foods have been all too easily dismissed as emotional, anti-science and anti-progress. There is clearly a mismatch between the government's attitude and most of the public's, and a failure to fully appreciate the reasons for this concern. It is too simplistic, too convenient and undemocratic merely to dismiss consumer concerns as ill-informed and non-scientific.

New research by CA undertaken in May 2002[1] shows

- There is widespread confusion with just over a third aware that there are GM foods on sale
- Consumers still feel very strongly about GM
- Less than a third (32 per cent) find the idea of food produced from a GM plant acceptable
- Under a third find the use of GM bacteria (for example to produce vitamins) or yeast acceptable
- Just 11 per cent find GM animals acceptable and 13 per cent, GM fish
- Half of respondents thought that GM could offer benefits for food production, but at the moment those developing the technology are seen to benefit most from it – with GM seen as offering least benefits for consumers
- Overall 57 per cent had concerns about the use of GM in food production. The main concern was not knowing enough about it and fears about tampering with nature. The research also showed great reluctance to accept growing of GM crops. Just 32 per cent were in favour of growing GM crops for commercial purposes in the UK at the present time. The main reason why people didn't think it should go ahead was again concern about lack of information

At the moment, consumers still do not have an adequate choice over the products that are already on the market. They do not have the means to completely avoid GM products if they want to, although some progress has been made by the main retailers, manufacturers and caterers to remove GM ingredients. Results of a CA survey of retailers, manufacturers and caterers show that in most cases, this extends to GM ingredients and to derivatives (such as soya oil for example). However, some are just basing their systems on those ingredients that are detectable in the end product, a limited number are trying to exclude GM ingredients from animal feed and very few include processing aids as part of their policies. None felt able to make GM-free claims, reflecting how widespread the use of GM ingredients and processes is already, and the difficulty of tracing and totally excluding GM. This problem is likely to become more difficult as more crops are grown world-wide and particularly if GM crops are grown in the UK and elsewhere in Europe.

Although GM has the potential to offer consumer benefits in the future, including improved nutritional content and taste for example, this is not evident from the crops currently on the market which have been modified to offer benefits to producers, such as tolerance to weed killers.

Controls

The EU is currently reviewing labelling rules in order to include GM-derivatives that are no longer detectable in the end product, based on traceability. Safety controls are also being reviewed in order to strengthen the approval process and give greater responsibility to the newly created European Food Safety Authority (EFSA). If they are adopted, these measures, which include for example provisions for long-term monitoring, approval for a limited period and greater importance given to risks and benefits as part of the approval process, will give consumers greater protection, but several issues will remain unresolved. These include the lack of any approval process for the many processing aids already widely used in food production, the need to incorporate more effective mechanisms for identifying unintended effects of genetic modification into the process and methods for monitoring long-term health effects.

Apart from ingredients to be used in foods, the issue of whether or not GM crops should be grown in Europe, and more specifically in the UK, is going to be a crucial issue over the next year. Farm scale evaluations of herbicide-tolerant crops (resistant to weed killers) are taking place in the UK. The results will be evaluated in 2003 and feed into a decision on whether or not commercial growing should take place. An assessment of the costs and benefits will also be undertaken by the Prime Minister's strategy unit.

As well as the need to reassure consumers that the potential for any health and environmental effects have been dealt with, the possibility of co-existence with conventional and organic crops also needs to be addressed, so that those who wish to avoid GM can do so.

A global approach

Part of the solution should be a common international approach to the traceability, labelling and safety assessment of GM food products. However, the body that sets these standards is making slow progress due to the different approaches adopted to GM regulation around the world which has made reaching agreement difficult. At the moment, there is too much focus on trade and producer interests at the international level. GM developments have also highlighted the influence that the World Trade Organisation (WTO) now has over the level of protection consumers can expect at national level.

Conclusion

Our position remains that GM has the potential to offer consumers benefits; however, we have yet to see such products and the health, environmental and ethical aspects raised by GM have yet to be adequately addressed. Consumer concerns should be addressed by regulators in the UK, EU and internationally as well as by industry, who dismiss consumer concerns at their peril. To do so, will only exacerbate consumer resistance to the technology.

Overarching recommendations

- There must be an effective and wide-ranging public debate about the future of GM foods and the conditions and limitations of public acceptance. The results must be reflected in future government policy, including any decision on whether to grow GM crops commercially.
- There should be no further GM products coming onto the market and no commercial planting of GM crops in the UK until the following issues have been addressed:
- An effective and meaningful public debate
- Effective consideration of consumer concerns, including the potential risks and benefits of a particular application as a part of the approval process
- More open, transparent and inclusive regulatory processes
- Mechanisms for monitoring the long-term consequences of GM for human health and the environment
- More independent research into the long-term consequences of GM
- Better mechanisms for picking up unintended effects as a result of the modification
- Full traceability of GMOs in place in order to track GM developments and have knowledge about where GM ingredients are used
- GM ingredients must be properly labelled based on what is used rather than what is detectable in

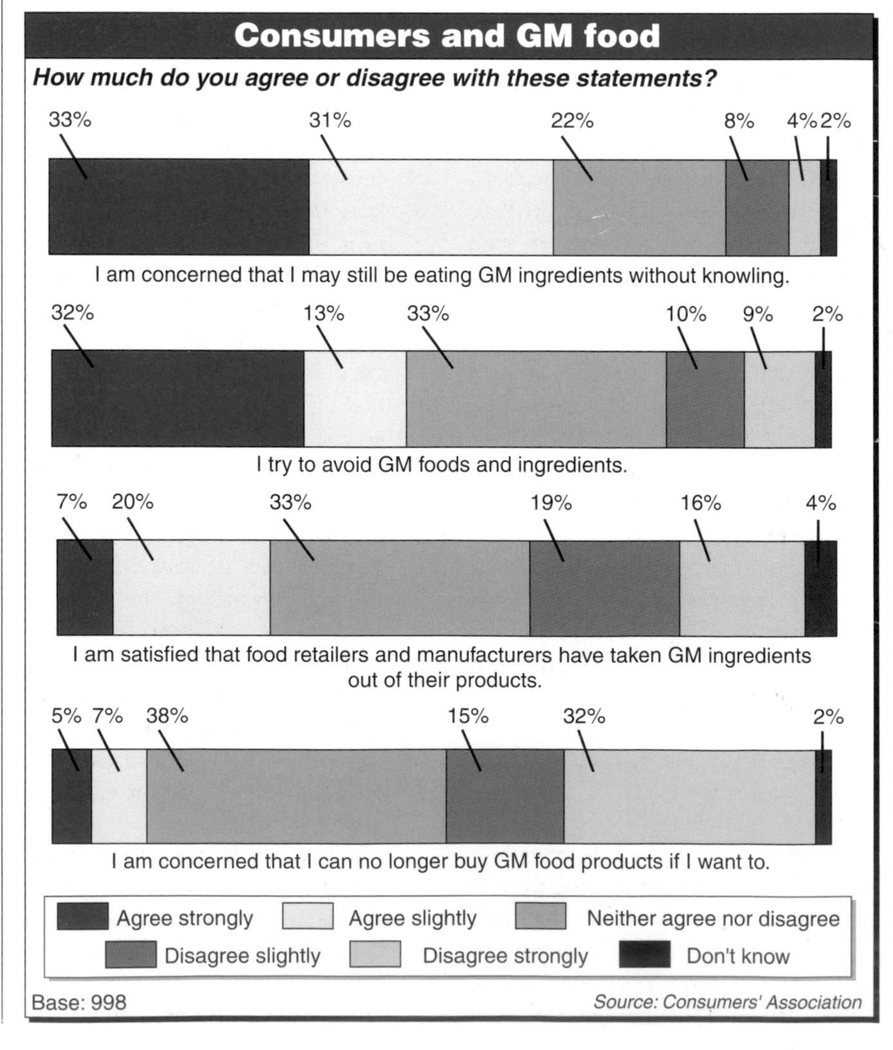

the end product, alternatives to GM products must be available and all labelling rules must be effectively enforced
- Globalisation demands an international approach to control of GM foods – organisations developing international standard must make progress and ensure that consumer concerns are recognised and addressed.
- For more specific recommendations please see the full policy report available at www.which.net/campaigns/gm

Reference

1 Questions were placed on BMRB's ACCESS Face-to-Face Omnibus survey during the period 16-22 May 2002. This is a weekly survey providing 998 in-home interviews with a nationally representative sample of adults aged over 15 across Great Britain. The final results of the survey are weighted so that they are in line with the national demographic profile.

■ The above information is from *GM dilemmas – Consumers and genetically modified foods*, by the Consumers' Association. For more information visit their website at www.which.net

Benefits

Information from CropGen

What are the benefits of GM crops?

GM technology can help to reduce the amount of chemicals used in agriculture. For example, making crops naturally resistant to a specific pest means that fewer insecticides are needed to get rid of that pest with little impact on beneficial insects. In the same way, if crops are made resistant to a particular herbicide, just one or two applications will be enough to control weeds in the field rather than having to use a number of different herbicides. (Currently, without the use of chemicals, we would lose about 40% of our crops.)

In the future, GM crops will be able to provide foods with improved nutritional value, longer shelf-life and lower prices.

What are the real benefits of GM technology? Who is benefiting and who will benefit?

We will all benefit now and in the future: from better and cheaper food and other plant products, an environment less damaged by agriculture, more wild and recreational land saved from the plough, and less poverty and food shortages in developing countries.

Will GM technology benefit our lives and how? What's in it for me?

Direct benefits for the UK consumer in the form of products on the shelves will take time to come through but if we say 'no' to further development we will never have them, while more far-sighted people will benefit.

There are three primary areas of benefit: health, environment and the economy. For the UK in the short term we will enjoy:

- farmers able to use the best of modern methods, increasing their efficiencies and incomes, contributing more to society and to the economy, and being less of a burden on the tax base;
- consumers being offered the best quality products from our own country and around the world;
- in our own UK environment, the deployment of more 'environmentally-friendly' techniques without loss of productivity or the need to bring more land into agriculture, taking it away from recreational and wilderness uses;
- maintenance and improvements in our skills base, slowing down or preventing the loss of our scientists to more progressive countries;
- retention in the UK of extremely important technology-based industries based upon modern biology; these industries are already finding a better research and business climate abroad.

Will GM technology make life easier/give us better food/more nutritious or healthier food/food with a longer shelf life? Will it be cheaper (by how much and why?) or cost more?

All of those things, but not next week. New crops need time to be developed and properly tested; it takes time, effort, money, and cannot be rushed. Such food is likely to be less expensive than conventional alternatives and much cheaper than 'organic' food.

If we in the UK do not put our heads in the sand, we can expect benefits to begin to flow progressively within a few years. In the USA and Canada, the benefits are already to be had: lower use of pesticides, higher yields and better agronomic practice. Meantime, parts of the Developing World with fundamental food problems are now seeing clear benefits: crops with the potential to solve critical nutritional deficiencies, crops allowing farmers to improve their agriculture beyond mere subsistence and getting a little money in their pockets to begin to move out of poverty. Is that not worth having?

What is the biggest advantage GM crops can bring the world?

Adequate food supplies for a burgeoning population without excessive invasion of current non-agricultural land.

■ The above information is from CropGen's website which can be found at www.cropgen.org

Issues of concern

Information from the World Health Organization

These questions and answers have been prepared by WHO in response to questions and concerns by a number of WHO Member State Governments with regard to the nature and safety of genetically modified food.

How are the potential risks to human health determined?

The safety assessment of GM foods generally investigates: (a) direct health effects (toxicity); (b) tendencies to provoke allergic reaction (allergenicity); (c) specific components thought to have nutritional or toxic properties; (d) the stability of the inserted gene; (e) nutritional effects associated with genetic modification; and (f) any unintended effects which could result from the gene insertion.

What are the main issues of concern for human health?

While theoretical discussions have covered a broad range of aspects, the three main issues debated are tendencies to provoke allergic reaction (allergenicity), gene transfer and outcrossing.

Allergenicity

As a matter of principle, the transfer of genes from commonly allergenic foods is discouraged unless it can be demonstrated that the protein product of the transferred gene is not allergenic. While traditionally developed foods are not generally tested for allergenicity, protocols for tests for GM foods have been evaluated by the Food and Agriculture Organization of the United Nations (FAO) and WHO. No allergic effects have been found relative to GM foods currently on the market.

Gene transfer

Gene transfer from GM foods to cells of the body or to bacteria in the gastrointestinal tract would cause concern if the transferred genetic material adversely affects human health. This would be particularly relevant if antibiotic resistance genes, used in creating GMOs, were to be transferred. Although the probability of transfer is low, the use of technology without antibiotic resistance genes has been encouraged by a recent FAO/WHO expert panel.

Outcrossing

The movement of genes from GM plants into conventional crops or related species in the wild (referred to as 'outcrossing'), as well as the mixing of crops derived from conventional seeds with those grown using GM crops, may have an indirect effect on food safety and food security. This risk is real, as was shown when traces of a maize type which was only approved for feed use appeared in maize products for human consumption in the United States of America. Several countries have adopted strategies to reduce mixing, including a clear separation of the fields within which GM crops and conventional crops are grown.

Feasibility and methods for post-marketing monitoring of GM food products, for the continued surveillance of the safety of GM food products, are under discussion.

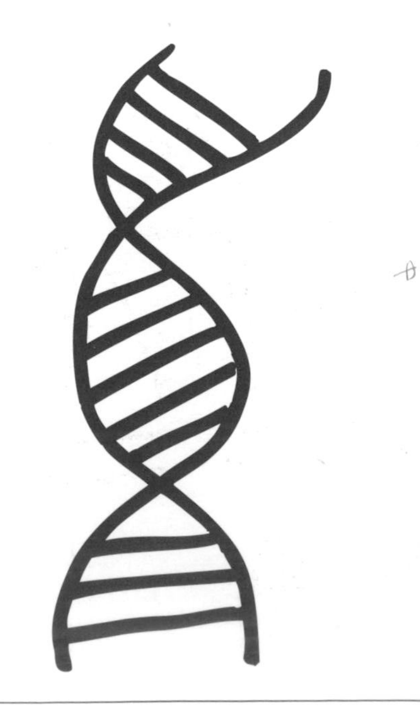

How is a risk assessment for the environment performed?

Environmental risk assessments cover both the GMO concerned and the potential receiving environment. The assessment process includes evaluation of the characteristics of the GMO and its effect and stability in the environment, combined with ecological characteristics of the environment in which the introduction will take place. The assessment also includes unintended effects which could result from the insertion of the new gene.

What are the issues of concern for the environment?

Issues of concern include: the capability of the GMO to escape and potentially introduce the engineered genes into wild populations; the persistence of the gene after the GMO has been harvested; the susceptibility of non-target organisms (e.g. insects which are not pests) to the gene product; the stability of the gene; the reduction in the spectrum of other plants including loss of biodiversity; and increased use of chemicals in agriculture. The environmental safety aspects of GM crops vary considerably according to local conditions.

Current investigations focus on: the potentially detrimental effect on beneficial insects or a faster induction of resistant insects; the potential generation of new plant pathogens; the potential detrimental consequences for plant biodiversity and wildlife, and a decreased use of the important practice of crop rotation in certain local situations; and the movement of herbicide resistance genes to other plants.

■ The above information is an extract from *20 questions on genetically modified (GM) foods*, produced by the World Health Organization (WHO). For the full document visit their website at www.who.int

Genetically modified organisms

Information for consumers

You can see foods in the shops labelled 'contains GMOs'. This is to give you a choice.

Genetically modified organisms (GMOs) are created by scientists in the laboratory by taking genes from one species and adding them to another. For example, genes may be taken from bacteria and added to a plant. It is the same idea as cross-breeding plants to get the colour of flowers you want. But with GMOs it is quicker and more accurate as it allows just one individual gene, or a small number of genes, to be inserted into a plant (or animal) to change it in a specific way. Most notably, it allows gene transfer between different species that couldn't possibly interbreed using traditional techniques.

Are GM foods safe?

When genetic modification (GM) is used in food production, each product has to be assessed for safety before it can be sold anywhere in the EU.

Concerns about GMOs mainly relate to their potential impact on the environment. Some GMO crops have been made tolerant to certain herbicides to assist in weed control. A worry is that if they interbreed with weeds, we could get super-resistant weeds. Others are made resistant to specific (insects) pests so that crop damage is reduced. The worry there is that there may be effects on other insects and on species that feed on insects. In theory at least, GM techniques could be used to improve the environment.

> *When genetic modification (GM) is used in food production, each product has to be assessed for safety before it can be sold anywhere in the EU*

If you want to spot GM food:

Under EU law, the presence of GM has to be labelled as GM, as long as it can be detected in the final product. The two main GM crops we may be eating are soya and maize (corn). Soya and maize derivatives are found in around 80% of processed foods.

GM soya and maize derivatives such as proteins and flour have to be labelled as GM.

Other derivatives obtained from soya and maize do not have to be labelled. They include soya or maize oil, starch, emulsifier, lecithin, glucose, fructose, dextrose, mono and diglycerides, malodextrin and sorbitol.

GM may also be used in our food chain because GM crops can be fed to farm animals like chickens, pigs and cows. There is no requirement to label foods from animals fed on GM crops because of the difficulty of distinguishing them from foods derived from animals fed on conventional feed. The idea of tracing GMs along the food chain is being considered by the European Commission but has not yet been adopted.

■ The above information is from Foodaware: the consumers' food group. For further information you can e-mail them at info@foodaware.org.uk or alternatively visit their website: www.foodaware.org.uk

Regulations

Information from the Food and Drink Federation

Consumer acceptance of the technology very much depends on confidence in the regulatory process. Many people do not realise that the regulatory hurdles which must be overcome before a GM variety is commercialised are daunting, and much more stringent than for conventionally bred varieties.

Any research, growing, import or use of GM crops is covered by EU-wide legislation. Regulations have been in place for more than a decade and are still evolving.

Tests in the laboratory

Any new genetic modifications must first be thoroughly tested in the laboratory, often for a number of years, to check that the proposed modification works with no adverse consequences. This is regulated by an EU Directive, which is implemented in the UK as the Genetically Modified Organisms (Contained Use) Regulation.

Research organisations and companies have to register with the Health and Safety Executive (HSE) if they wish to carry out such tests. Certain tests are subject to a detailed notification which is reviewed by HSE, as well as a number of Government departments and advisory committees.

Tests in the environment

If the genetically modified organism (GMO) is a plant variety, the GM crops must then be tested in the field. Both field trials and commercial growing are regulated under the EU 'Deliberate Release' Directive (Directive 2001/18 which updates the earlier Directive 90/220/EEC).

Research organisations and companies planning to carry out field trials in the UK must submit a detailed application to the Department for Environment, Food and Rural Affairs (DEFRA). This is comprehensively reviewed by the independent Advisory Committee on Releases to the Environment

(ACRE), to assess possible risks to the environment. If further scientific information is needed, ACRE will order research to be carried out. For example, ACRE has requested additional research into the potential risks of GM insect resistant crops to beneficial insects. Members of ACRE include experts in ecology, biodiversity and farming practice.

The next step is to grow and market the GM crop on a commercial scale. This requires an application to market the GM plant, which is reviewed both at national level and by the other EU Member States. In the UK, applications are reviewed by ACRE, in consultation with other advisory committees, which then provides advice to DEFRA. So far, clearance for commercial growing in the UK has been granted for one

GM crop, a herbicide tolerant maize for animal feed. The company which developed this variety has decided not to grow it in the UK for some time. It is unlikely that any GM crop will be grown here commercially until 2006 at the earliest. Other crops have been approved for growing in the EU, such as herbicide tolerant oilseed rape.

Farm scale evaluations

In the UK, a series of Government-sponsored Farm Scale Evaluations, begun in 2000, compared the environmental effect of growing GM herbicide tolerant oilseed rape, maize and sugar beet against conventional varieties. They assessed the effect of the associated herbicide use on a range of plants and insects and other wildlife by comparing the diversity and abundance of numbers of key species during three growing seasons. The results of the trials, published in October 2003, have been passed to ACRE and will help to guide the Government on decisions about future commercial growing.

The trials were developed and overseen by an independent board, including members from organisations such as English Nature and the RSPB. A cross-industry body, the Supply Chain Initiative on Modified Agricultural Crops (SCIMAC), was established to support the open and responsible introduction of GM crops in the UK. It has played a major role in finding suitable trial sites and in developing best practice guidelines. SCIMAC may play a longer-term stewardship role, addressing both the management of GM crops and co-existence with non-GM crops.

These trials were not set up to assess the safety of growing these crops. The regulatory authorities have already assessed this separately on the basis of significant experimental work, and have no concerns about the safety of the crops grown under the specified conditions. Nevertheless, since these are not commercial plantings, the crops are destroyed after harvesting, and hence do not enter the food chain.

These trials have been quite controversial. In some areas, there has been significant local opposition, and a number of incidents of people trying to destroy the crops. In other places, the local community has voiced little concern.

Supporters of the technology point out that globally, 25,000 field trials of GM crops have been carried out so far with no significant adverse consequences and it is only through field trials, and then larger farm-scale trials, that the safety or otherwise of the technology can be determined. Others take the view that our fragile environment is too precious to be put at risk, however remote that risk may be.

Use in food

In the UK, the Food Safety Act requires that all food, however produced and processed, must be safe, fit for consumption and free from contamination.

Specific safeguards control the introduction of 'novel foods' into EU member states. This term refers to foods or food ingredients which have not previously been used for human consumption 'to a significant degree' within the EU. New EU Regulations were introduced in April 2004 to control the safety assessment and labelling of foods consisting of, containing or derived from a GMO. These Regulations replace the GM element of the 1997 Regulation on Novel Foods and Novel Food Ingredients which now applies only to other forms of 'novelty' and not to GM foods.

In the UK, all applications for novel foods, including foods consisting of or containing GMOs or their derivatives, are assessed for safety by the Advisory Committee on Novel Foods and Processes (ACNFP) and other independent advisory committees. These bodies are made up of independent experts, including consumer representatives. The Food Standards Agency and DEFRA consider this advice as part of their statutory role in the approval of foods consisting of, or derived from, GMOs. This responsibility will continue under the new GM food and feed Regulation, though the responsibility for assessing the GMO's safety for the EU as a whole will rest with the European Food Safety Authority (EFSA).

In the UK, the Food Safety Act requires that all food must be safe, fit for consumption and free from contamination

Labelling and traceability

Labelling is an important way of informing consumers about foods, including the use of ingredients, which have been genetically modified. The new GM food and feed Regulation requires that all foods consisting of, containing or derived from GMOs must be labelled as such. This marks a major change from previous legislation which required food and food ingredients to be labelled only if novel DNA or protein was detectable. For example, highly refined products such as oil or glucose syrup from a GM source did not require labelling under the Novel Foods Regulation, as they contain no modified DNA or protein. However, these products do now require labelling under the new GM food and feed Regulation.

Identity-preserved ingredients

Many consumers want to avoid GM-derived ingredients. In response, many manufacturers and retailers have established supply lines where the conventional crop is kept separate from the GM crop. This non-GM supply is referred to as 'identity-preserved'. FDF, working closely with the British Retail Consortium (BRC), developed and disseminated the internationally recognised BRC/FDF Technical Standard for the Supply of Identity-Preserved Non-Genetically Modified Food Ingredients and Products.

In practice, it is extremely difficult to avoid some co-mingling of GM material with conventional supplies in a supply chain which covers harvesting, storage and transportation of bulk supplies. The law therefore allows non-GM crops to contain up to 0.9% GM material, as long as the non-GM origin of the crop can be clearly traced and appropriate systems to maintain a separate supply chain are in place. These standards are not intended to support 'GM free' claims.

GM in animal feed

Some people do not wish to buy foods produced from animals which have been fed on GM feed. However, the GM food and feed Regulations do not require food labels to indicate if a food has been derived from animals fed on GM feed.

European regulations

There have been no approvals of GM crops in the EU for several years while the new legislation to integrate the whole food supply chain, including traceability, was in development. The first of the new Regulations establishes procedures for tracing, labelling and maintaining records of GMOs right along the supply chain. It covers all products in which GMOs are used, not just food and feed, so could include jeans and T-shirts produced from GM cotton.

The second Regulation (GM food and feed) lays down detailed rules for the approval and marketing of GM crops and GM-derived food or feed. It covers safety assessment, labelling and provides for the establishment of a register of authorised GM food and feed.

Both Regulations cover GM feed, as well as food, to ensure that the authorisation encompasses all possible uses of the GM crop. The Regulations take GM foods outside the scope of the Novel Foods Regulation and provide a stringent set of controls specifically for GM foods.

Once these Regulations were in place, the first new approval, for the marketing of a GM variety of a canned sweetcorn, was granted by the European Commission in May 2004.

■ foodfuture is produced by the Food and Drink Federation on behalf of the UK food and drink manufacturing industry. For more information visit the website www.foodfuture.org.uk

GM food – opening up the debate

Information from the Food Standards Agency

Is GM food on sale in the UK?

Three GM foods or ingredients have been on sale or approved for use in foods in the UK:

- GM tomatoes, which were sold only as tomato purée
- GM soya
- GM maize

No fresh GM produce has been approved for sale or consumption in the UK. There have been no animal, fish or human genes approved for use in GM food anywhere in the world.

However, many processed foods in the UK, such as biscuits, cooking sauces, and food coatings, will include GM ingredients at a very low level if they use soya or maize as an ingredient. The same will be true for products imported from countries growing GM soya or maize. So, unless an individual's diet contains no processed foods, they are likely to be eating at least some GM or 'GM-derived ' food, even if this is only at a low level. If they have travelled to one of the countries that grow GM crops in the past few years, especially the USA and Canada, it is very likely that they will have eaten food that contains GM material or is derived from it.

Animal feed

Also, as more countries have started using genetic modification technology, supplies of animal feed ingredients to the UK have increasingly contained GM or 'GM-derived' products. Maize, soya and oilseed products are major sources of energy and protein for UK livestock. Imported soya and maize by-products account for around 20% of the raw materials used by UK feed manufacturers and farmers.[1]

Many processed foods in the UK, such as biscuits, cooking sauces, and food coatings, will include GM ingredients at a very low level if they use soya or maize as an ingredient

More than half the soya bean crop grown worldwide is GM, and increasing proportions of maize and oilseed rape crops are GM.[2] Some animal feed is therefore likely to contain ingredients derived from GM crops, but it is not possible to say how much because the exporting countries do not routinely separate their GM and non-GM crops. As an indication, however, 30% of the US maize crop is GM, while 75% of US soya and 95% of Argentine soya planted in 2002 were GM.[3] Some animal products, including meat, milk and eggs, therefore come from animals that have eaten GM feed, either here in the UK or in their country of origin.

The amount of animal feed consumed by different types of livestock varies according to their diet. Cattle and sheep eat a higher proportion of grass in their diets, as opposed to pigs and poultry, which eat more maize and cereals.

However, the diets of all species of livestock are supplemented by manufactured compound feeds,

particularly during the winter. The choice of what feed to use is determined by factors such as the age and species of the animal and the availability and price of feeds.

How much GM food is on sale in Europe?

Currently, food ingredients from varieties of GM soya, maize and oilseed rape have been approved for food use in the European Union although very little is actually used. These include oils and syrups that contain 'GM-derived' material, and flours and starches. These ingredients *could* be used in a wide range of processed foods, from vegetarian burgers to biscuits and sauces, in the same way ingredients from non-GM crops are used.

However, many food manufacturers and supermarkets have said that they are excluding ingredients from GM crops from their products. This started in the late 1990s, when people became more concerned about genetic modification.

Some GM products, such as chymosin, may be used in food manufacturing as 'processing aids'.

Will the label tell me if the food is GM?

In the EU, if a food contains GM DNA or protein, this must be indicated on the label. The same applies when eating out. If any food or drink contains GM DNA or protein, then a notice indicating this should be displayed.

The law on labelling of GM foods is based on the ability to measure differences in composition between GM and non-GM ingredients. On this basis, certain ingredients do not need to be labelled. For example, refined vegetable oil because this has exactly the same composition as oil obtained from non-GM crops and is therefore indistinguishable from it. Any intentional use of GM ingredients at any level must be labelled. However, there is no need for small amounts of GM ingredients that are accidentally present in a food to be labelled, if they make up less than 1% of the product.

The current labelling rules are, however, being extended to cover all food and animal feed that contain any material that comes from GM sources, whether or not any GM material is present in the final product. This includes products such as oils but not food made with the help of genetic modification technology, such as hard cheese. These products, and products such as meat and milk from animals fed on GM feed, will not need to be labelled.

The new rules, which were agreed in December 2002 and are awaiting implementation, are compared with the current rules in the table.

Labelling rules in the UK

Categories	Examples	Current situation	New rules
1. Products made from GM crops, but no GM material present	Highly refined maize oil, soya bean oil, rape seed oil, alcoholic beverages	Labelling not required	Labelling required
2. Products made from GM crops and GM material present	Maize, soya bean sprouts, tomato, maize flour	Labelling required	As current
3. Labelling of food made from genetic modification technology	Cheese produced with the help of chymosin from GM micro-organisms	Labelling not required	As current
4. Labelling of food made from animals fed on GM animal feed	Milk, meat and eggs	Labelling not required	As current
5. Threshold for approved GM varieties	Approved maize and soya	Threshold is set at 1%	Threshold is agreed at 0.9% in December 2002 but may be subject to change pending discussions in the European Parliament. There will also be a threshold of 0.5% for non-EU approved varieties
6. Food sold in catering outlets	Restaurant menus and takeaways	Optional (compulsory in UK where GM material is present in final food in line with current labelling rules)	Labelling required although detailed rules yet to be developed

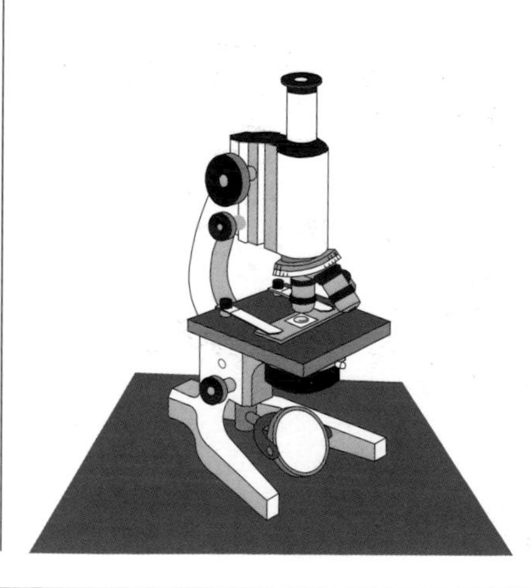

References

1 Estimates from the Grain and Feed Trade Association
2 James, C. *Global Status of Commercialized Transgenic Crops: 2002*. ISAAA Briefs No. 27: Preview. Highlights of this report can be seen on the Crop Biotech Net site at: www.isaaa.org/kc
3 Sources: Food Standards Agency, Grain and Feed Trade Association, US Department of Agriculture and American Soybean Association

■ The above information is from *GM food – opening up the debate* produced by the Food Standards Agency. For further information see their website at www.foodstandards.gov.uk

Why it's time for GM Britain

Leaked papers show the cabinet has given the go-ahead to genetically modified maize. Top government science adviser Chris Pollock puts the economic and ecological case to Ian Sample

The back streets of Aberystwyth are narrow, windy and perfect for Chris Pollock's cramped little MGF convertible. He's giving me a lift to the railway station, with a taxi driver's sense for traffic-dodging short cuts, throwing the car down unpromising-looking streets, jumping ahead of the queues of Ford Focuses and Vauxhall Corsas.

Pollock is a big man to be driving such a small car, but then he's a big man for Aberystwyth. He is director of the Institute of Grassland and Environmental Research, the government-funded research centre for science in agriculture and the environment. He also heads the government's advisory committee on releases to the environment (Acre), the scientific panel that briefs ministers on the risks of genetically modified organisms in the environment. The position puts him at the frontline of one of the most fraught scientific debates the country has faced in recent years.

Recently, that debate has taken another turn. If the leaked minutes of a recent cabinet meeting are accurate, the government intends to follow advice from Acre and announce the go-ahead for a variety of GM maize, for the first time opening the door to a GM Britain.

If today's GM crops don't provide overwhelming advantages over conventional crops, the technology is necessary not just to improve agriculture, but to stay competitive

Pollock is passionate, talking in booming sentences, about the stuff outside his office's huge windows – acres and acres of countryside, today looking washed out after heavy rain. As a self-confessed technophile, he believes it is right for GM crops to play a role in Britain's future agriculture. 'There's no good scientific reason, in my view, for saying GM crops will innately cause problems purely because they are genetically modified,' he says. 'As a technophile, my argument would be that ever since the invention of the internal combustion engine, we've managed technology, with mistakes, but we've managed it to the extent that life expectancy continues to go up at a year a decade.'

But does the view that GM crops, if managed responsibly, will not cause problems justify their introduction? Aren't GM crops technology for technology's sake, at least in Britain where, unlike in some developing countries, crops are rarely wiped out by aggressive pests or harsh weather, or simply fail to thrive in the soil? Pollock argues a longer-term view, that even if today's GM crops don't provide overwhelming advantages over conventional crops, the technology is necessary not just to improve agriculture, but to stay competitive. 'If the UK is going to play some part in a global agricultural market, any new technology that reduces the price of a crop will have to be taken up if you are not to be at a disadvantage. Developed countries

need to implement new technologies to stay ahead of the game,' he says. Despite the Cabinet Office's surprisingly downbeat report last year, which claimed, among other concerns, that consumers would see little benefit from the current, first generation of GM crops, Pollock believes that the public will ultimately be better off if they are introduced. 'Historically, every time you do something more efficiently, what happens is the price falls. The profit does not stay with the farmer. You can make the same argument about integrated circuits. Chip manufacturers go out of business like there's no tomorrow because every time they make them better, the price goes down and the benefit gets shoved right down to the consumer,' he says.

And it is everyday consumers who will decide the future of GM crops in Britain. If GM crops flop in the shops, only a foolish farmer would continue to churn them out. 'If nobody wants to buy it, there's no advantage to growing it in the UK. Farmers are highly responsive to markets. If they weren't we'd have a horsemeat industry in the country even though no one eats horse. These guys have got to make money out of it,' says Pollock.

Aside from the ongoing, still stubbornly polarised debate on GM crops in Britain, there are deeper concerns about British agriculture that have largely gone unaddressed. According to Pollock, by focusing attention on the impact of farming techniques on the environment, the government's farm-scale trials brought to the fore questions about the impact intensive farming has already had on Britain's wildlife and broader environment.

Britain's push towards more intensive farming was driven by postwar necessity: in 1945, there was simply too little food to go round. There was no motivation to think about, let alone squabble over, the best kind of agriculture for the country: in short, it was no time for the luxuries of caution and complaint. As a result, Britain leapt into a regime of intensive farming on a massive scale, creating a legacy from which we are still suffering. 'Since the war, we have destroyed habitat diversity, we've destroyed habitat quality, simply because there wasn't enough food to go round and we needed to increase crop yields,' says Pollock. 'The country swallowed whatever agricultural technology was thrown at it.'

Now, says Pollock, it's time we decided what we really want from our countryside. Change in land use is inevitable, driven in part by the forthcoming, though undoubtedly sluggish, reform of the common agricultural policy that will see subsidies for farmland change dramatically.

It is everyday consumers who will decide the future of GM crops in Britain. If GM crops flop in the shops, only a foolish farmer would continue to churn them out

Pollock sees it as an opportunity to undo some of the harm farming has caused in recent decades. Intensively farmed land could be converted into tourist-attracting grounds that are managed to maintain their biodiversity. 'Most people will accept that we've gone too far and we need to reel back. The point is, all this land has to be managed. You can't just leave it untended and expect it to turn into a zoo without your doing anything.'

But how far back do we want to rewind the British countryside? What balance of biodiversity do we want? These are the questions the public should be thinking about now, he says.

Unsurprisingly, Pollock says science, and GM technology, may provide novel ways of changing the countryside. If the technology works as its advocates hope, GM crops might allow farmers to produce the same yields from smaller acreages, leaving more land aside for wildlife. In practice, that means a mix of more hedgerows, so-called shelter belts, broad field margins and strips within crop fields where creatures can seek food and refuge.

Much of the land currently being farmed will be given up for other uses. 'If you have shedloads of money, what's the most expensive commodity you can buy in Britain?' asks Pollock. 'It's privacy.' Those with the money to do so are increasingly buying up land, not to build flats on or turn into some other eyesore, but to have as their own, to walk around in, to plant trees in, to enjoy for what it is. 'If you look at land sales in the south-east over the past six months, it hasn't been going to farmers or property developers, it's going to stockbrokers,' he says.

Pollock admits that he has succumbed to the lure of land-buying himself, having bought a small patch nearby to walk around in.

While Pollock is keen to move on, to think about other issues of import to the British countryside other than just GM crops, he remains palpably ruffled by Britain's tortuous GM debate. Scientists should take some of the blame, he says, for being unprepared for the public's reaction to the powerful new tools to spin out of biotech research. 'The scientific community got very badly flat-footed over GM. I don't think they thought there would be a problem,' he says. 'They just didn't realise the world was going to change – to make public engagement so important – so quickly.'

The result was that when non-governmental organisations began to voice concerns, the scientific community didn't know how to react. Part of the difficulty was that many lobby groups were attacking GM on different, single issues, such as the potential impact on birds or the impact on organic farmers. Pollock argues that the environment cannot be treated that simply. A rational debate has to address all the issues, he says. But that is not the way it happened, and effective, headline-grabbing stories became the norm. 'You get messages like "don't eat GM food because it will kill you", which is palpably crackers, from people who smoke marijuana,' he says.

According to Pollock, the wrangling over GM crops, and the demonising of GM technology as a whole by some, has had a lasting effect on the public's view of scientists working on GM crops. If they are advocates of the technology, they are necessarily in the pockets of the biotech industry, the government or both. Or so the anti-GM lobby groups would say.

'We've gone from the days of Harold Wilson and the white-hot heat of the technological revolution to scientists like me being labelled next to the proprietor of the *Sun* as the people least trusted by the public,' says Pollock. 'There was a guy on the television the other night talking about how they were producing a vaccine for bird flu using genetic manipulation. If I said I was genetically modifying a chicken so it couldn't catch bird flu, I'd be hanging from a lamppost by now. It's been an uphill struggle with GM, I don't think there's any doubt about that.'

Continuing opposition to GM foods

The British public is still strongly opposed to genetically modified (GM) foods, according to new research by the MORI Social Research Institute.

The Government is entering a consultation process to decide how the UK responds to the controversial subject of GM food. It does so, however, with the public firmly opposed to the introduction of GM foods – more than half the public (56%) opposes GM food, compared to one in seven (14%) who support it.

Opposition is remarkably stable across all political perspectives; 56% of intending Labour voters are against GM, as are 57% of intending Conservative voters and 60% of Liberal Democrat voters. Women are more likely to be opposed than men (61% versus 51%), but there is little difference by household income. Opposition is not a luxury for the affluent; 56% of households with a gross annual income of less than £17,500 oppose GM, as do 60% of households with an income of more than £50,000.

However, while there are very few active supporters of GM, one in four (25%) are neutral. Therefore, public opinion is split three ways; the majority opposed, a significant minority neutral, and only a small minority advocates.

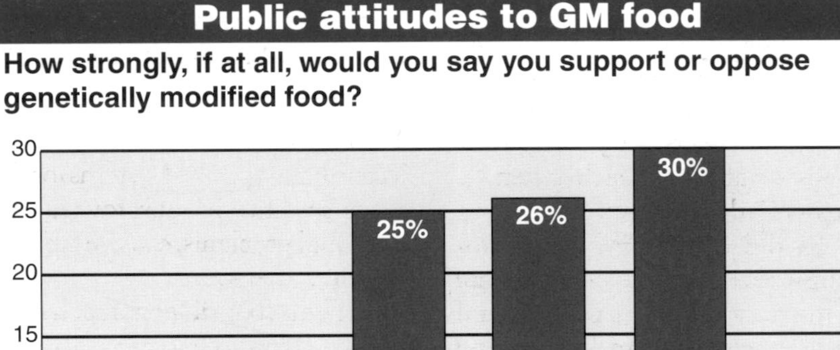
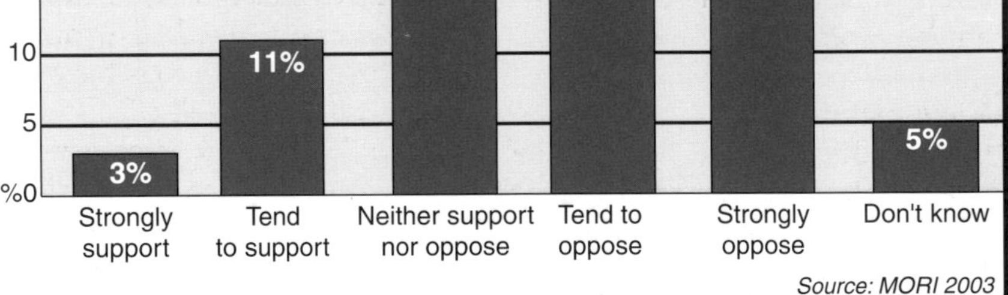

Philip Downing, head of environmental research at MORI Social Research Institute, notes: 'The public is yet to be persuaded about GM food and shows little sign of softening its opposition, which has been relatively stable ever since the issue came into the media spotlight.'

The result is part of MORI's own environment research, reported in full in the MORI Environment Research Bulletin. Other results include strong support for a 10p tax on supermarket plastic bags, huge support for renewables over nuclear power, and divided opinions on extending congestion charging to other UK towns and cities.

Technical details

The Environment Research Bulletin Survey was carried out among a representative quota sample of 2,141 adults aged 15 years +, face-to-face and in-home across 201 sampling points in Great Britain between 6-10 February 2003. All data have been weighted to the known profile of the British population. For a copy of the MORI Environment Bulletin visit the MORI website at www.mori.com/environment

■ The above information is from MORI's website which can be found at www.mori.com

Is GM food safe?

GM Nation?

Is GM good for me?

People disagree whether GM foods pose a health risk to those who eat them.

Views for

GM crops are tested much more extensively than non-GM crops.

Any differences between the GM and non-GM crops are studied in detail and full health and safety assessments are made. Such tests include food safety, nutritional quality and allergenicity. No crop or food can be grown or sold unless it is proven that they are at least as safe and nutritious as their non-GM counterparts.

Hundreds of millions of people and animals worldwide eat food and feed from GM crops and there have been no ill effects, supporting fully the responsible approach taken by industry and regulators.

In the future GM crops providing extra nutrients or medicinal substances will make a positive and direct contribution to human and animal health.

Views against

GM foods might be bad for our health. There has not been enough research to evaluate the short- or long-term effects of GM foods on both human and animal health.

The following information is from GM Nation? – The public debate. This was the UK's first nationwide public discussion around GM issues. Meetings were organised across the country by local authorities and network groups.

The risk assessments done so far are not robust enough to take into consideration all the unknowns and uncertainties that currently exist. We may be in for some unpleasant surprises.

GM foods aren't needed for a healthy, balanced diet.

Are GM foods safe?

There is disagreement about whether current GM foods can cause health problems.

Views for

Testing and approval systems for GM crops specifically look for real and potential problems, e.g. the creation of a new allergen.

The responsible approach taken by industry and regulatory authorities has meant that hundreds of millions of tonnes of GM crops have been eaten without any ill effects.

Indeed, the development of a soybean containing a Brazil nut protein was halted by the company involved at the research stage as soon as an allergen problem was detected, and before it was ever grown outside of a greenhouse. This demonstrates that the safety assessments and regulatory system work.

In 2002 an EU study which looked at 14 years of research by 400 groups of independent scientists concluded that 'GM crops are probably safer than their non-GM counterparts'.

Views against

Inserted genes might come from plants and animals that have never formed part of our diet before, and this might lead to unexpected health problems. Also, the insertion of the gene might disrupt normal function in ways we can't see or predict. In both cases, the new foods might produce substances that, for example, trigger allergic reactions.

Genes can cross from the food we eat to the bacteria in our stomachs, and these might include antibiotic-resistant genes that have been routinely used as marker genes in GM technology. This could render antibiotics ineffective against human and animal diseases.

Testing is not adequate and a full system of safety checks is needed. Safety testing systems cannot be 100 per cent effective in predicting allergenicity because we don't know why some things cause allergies and others don't.

What about allergies/ diseases?

People disagree whether GM foods present a different kind of threat to health than new non-GM foods.

Views for

This question is not specific to GM and can be asked of any new food that comes on the market.

All GM food is tested for its allergenic potential. There is no evidence at all of allergic and other reactions to the foods from the GM crops currently being grown and eaten by millions of people.

In 2002 an EU study which looked at 14 years of research by 400 groups of independent scientists concluded that GM crops are probably safer than their non-GM counterparts.

Future GM crops may in fact eliminate allergens from foods such as the peanut or wheat.

Views against

Because GM crops introduce new proteins into our diet, it is only a matter of time before a GM food will be shown to cause one or other disease or allergy.

There is already evidence of increasing allergy to soya products for example, which might be linked with the growth in the use of GM soya. We do not know enough about what triggers allergies to make our current testing procedures adequate. Long-term population studies are needed. So far, they have not been done, and claims that there have been no effects cannot be believed.

Will GM food be harmful in the future?

People disagree about the future risks of eating GM foods, and about whether these risks are justified.

Views for

This question is not specific to GM and can be asked of any new food that comes on the market.

Testing and monitoring systems are stringent and robust and GM crops and foods can only be grown and eaten if it is proven that they are at least as safe as their non-GM counterparts.

As far as eating DNA from GM

foods is concerned, we eat DNA every day, in our food. DNA is and always has been digested in our stomachs and DNA from GM plants is no different.

Views against

GM is being pushed forward far too fast, without adequate monitoring.

The current testing and monitoring regimes deal with what we know now and are inadequate to pick up long-term problems. We need new, more rigorous testing methods. Because we don't know enough about how genes work, there may be unpleasant surprises later down the line, after we may have already committed ourselves deeply to GM.

The problem is, harmful effects will probably only emerge from large-scale experiments in nature, which exposes us to the risk anyway.

Is altering genetic make-up safe?

People disagree about whether changing the genetic make-up of organisms through GM technology is harmful, or merely an extension of current breeding practices.

Views for

Mutation (changes to the DNA code) is happening all the time in nature. This is a fundamental process of evolution.

Conventional modern breeding uses chemicals and irradiation to create mutations and new varieties. GM is only different from these processes in that it is more precise. Testing regimes specifically watch out for any unexpected events triggered by the process and such events are eliminated.

Views against

GM makes it possible to take genes from one species and put them into a totally different species, crossing plant-animal boundaries for example. Effects of that transfer on the host organism are unpredictable.

Scientists do not know enough about how genes work together to claim that there is no harm. Some unexpected harmful results might be missed in testing or only appear later.

Can we cope?

People disagree about the potential for GM foods to cause problems and whether we have the ability to prevent or respond to them.

Views for

'Could we cope with any problems?' needs to be asked of all agriculture, not just GM-based systems, and of all foods.

There is no evidence that extra risks to health are involved in GM. In fact, non-food GM crops are being developed that will help treat and prevent disease. These include vaccines and medicines that can be produced more cheaply and more effectively in plants than in laboratories.

GM crops are subject to a great deal of monitoring, which will be able to pick up problems and ensure that they are dealt with.

Views against

We actually know very little about what is happening, and how difficult it is therefore to track what harm is being or might be done.

The safeguards in place are based on assumptions, not evidence, and on the current state of affairs not the future. If GM technology is accepted, our exposure to its products, in food and in the environment, will radically increase.

Any problems in living systems are different in kind from other products – once an organism is released, or cross-contamination occurs, we cannot 'recall' them if things go wrong.

■ The above information is from the Government's GM Public Debate Steering Board. For more information visit www.gmnation.org

What is being done about the potential risks?

Information from the Biotechnology and Biological Sciences Research Council (BBSRC)

For the environment

Genes from crops can pass into wild relatives of the crop. This will happen with both conventionally bred and GM crops. The extent of such gene flow and its significance depends on the crop: some have very few or no wild relatives in the UK.

GM crops could be made 'male sterile' so that their genes could no longer be spread through pollen. Or genes could be introduced in such a way that they are not passed on in pollen.

Spread of genes from GM crops that are resistant to a herbicide could create weeds that are resistant to it ('superweeds').

Some conventionally bred crop varieties and weeds are already herbicide-resistant. If GM crops led to herbicide-tolerant weeds they would be tolerant only to that herbicide and could be controlled by others if required. Many wild weeds in the countryside are never sprayed, so whether or not they contain a gene for herbicide tolerance does not affect them. Scientists are currently comparing the behaviour of a non-GM herbicide-tolerant oilseed rape with a non-GM, non-tolerant variety and a GM tolerant variety to determine optimal management strategies.

GM crops might grow as weeds (volunteers).

In UK trials, herbicide-tolerant GM oilseed rape does not appear to pose any more problems than conventionally bred varieties. Many crops are relatively poor competitors: wheat for example can only survive for about 2 years in the wild.

GM crops that are resistant to insect pests might deplete pest populations and so damage the natural food chain.

Any form of pest control e.g. mechanical, chemical, organic or GM has the potential to deplete pest numbers and so impact on the food chain. GM crops could be designed so that they switch on the gene that protects against pests only when the crop is under severe attack by the pest, or only when it is sprayed by a harmless compound. This would prevent depletion of pest populations.

GM crops that are resistant to insect pests might accelerate the evolution of pests to overcome this resistance.

Resistant pests evolve in response to any control method. Careful and limited use of spraying and using different sprays in rotation are among the strategies already employed by farmers to minimise this effect. With GM crops there is the potential to use two or more different genes that confer resistance against a pest, in tandem in a crop. This would make it much harder for pests to overcome the resistance because they would need to evolve two or more changes at the same time. Alternatively, farmers could rotate crops that each contain a different gene conferring resistance against the pest. Scientists are exploring the use of 'refuge areas' of non-GM plants in and around GM crops, in which populations of pests would be free from any pressure to evolve resistance.

GM crops that are resistant to insect pests might be harmful to other species.

Scientists are conducting laboratory and field trials to study the impact of insect-resistant GM crops on interactions between the plant, its pests and the pest's predators or parasites. The genes used in current GM insect-resistant crops code for natural compounds produced by the bacterium *Bacillus thuringiensis* (Bt) that are used in sprays approved for use in organic farming. These compounds are highly specific: for example, the agent against butterfly and moth larvae does not affect bees.

GM crops will reduce biodiversity

A preliminary trial on GM herbicide-tolerant sugar beet has shown that leaving weeds to grow for longer in the crop before spraying increases the number of insect species living on them; after spraying, there is a new range of species that colonise the dying weeds.

For the consumer

Fears have been expressed that GM foods might result in unexpected health hazards such as novel allergens, or transfer of antibiotic resistance from marker genes in GM crops to bacteria that live in the human gut.

Allergens can be introduced through conventional breeding. Tests are used routinely to detect and eliminate them. The same tests could be used with GM foods. GM technology could be used specifically to eliminate known allergens from foods.

Foods derived from GM crops may or may not contain any of the inserted gene or the protein for which it codes – it depends on the type of food. Technically it is becoming possible to produce GM crops in which the inserted gene would be active only in non-edible parts of the plants, where this would be appropriate.

Concerns about the long-term dietary effects of eating novel pieces of DNA apply equally to new varieties bred by conventional breeding. As all DNA is made up of the same four building blocks, and DNA sequences are broken down into short pieces of DNA by enzymes in the gut, it is inconceivable that novel combinations of these building blocks would be more likely to arise from GM than non-GM plants. Conventional, non-GM foods carry many unidentified microbial genes, which are consumed along with the genes of the plant or animal material.

Further research into the fate of DNA in the diet would be equally appropriate for GM and non-GM foods.

Alternative technologies have been, and are being developed to replace the use of antibiotic marker genes and eliminate unwanted marker genes from GM plants.

For the farmer

GM crops might show unexpected properties and might not breed true. Laboratory studies on a range of crops have shown that genes inserted by genetic modification into plants are stable, perform predictably and are inherited normally.

Just like conventional plant breeding, genetic modification may produce some offspring with unstable and undesirable characteristics. In both cases, these can be identified and eliminated from breeding lines.

Scientists have identified mechanisms by which genes inserted by genetic modification may very rarely get 'switched off'. They are developing options for more precise regulation of inserted genes.

■ The above information is from GM *agriculture in the UK?* published in 1999 by the Biotechnology and Biological Sciences Research Council's (BBSRC). For more information visit their website: www.bbsrc.ac.uk

Contamination threats

GM non-food crops will bring contamination threats to food and nature

A new GeneWatch UK report published April 2004 reveals how the production of GM crops intended for non-food uses could contaminate food crops and wild species. The 48-page, report 'Non-food crops: new dawn or false hope? Part 2: grasses, flowers, trees, fibre crops and industrial uses', was written by Dr Sue Mayer, GeneWatch's Director.

The report considers research taking place into the development of GM crops intended for non-food use: grasses, flowers, trees, crops such as cotton used for fibre production, and the range of different crops being modified to provide the raw materials for industrial production of oils, starches and plastics. It considers how they are being modified, how successful it has been and what environmental and health issues are raised. It makes recommendations for policy and research. The report reveals:

- how the biotechnology industry are pursuing the use of GM in non-food crops in the hope of sidestepping public concerns over GM foods;
- that GM grasses for use in lawns and golf courses may be commercialised in the US soon and raise serious environmental concerns because they are perennial, freely wind pollinating, and often spread via underground shoots. Grasses commonly spread worldwide on wool and via imported grass and bird seed;
- how GM trees for use in intensive plantations may pose serious environmental threats because trees are long lived and their seed and pollen can move long distances;
- that grasses, trees and fibre crops are often being modified in the same way as food crops including herbicide tolerance which has caused controversy over biodiversity impacts;
- the use of GM food crops, like oilseed rape, for non-food uses such as the production of biofuels and plastics, could lead to the contamination of non-GM and organic food crops;
- GM cotton and flowers are the first commercialised non-food GM crops that are being grown commercially outside Europe;
- GM potatoes modified for industrial starch production could be given approval for growing in Europe in 2004.

'The use of GM for non-food crops could bring contamination of food and nature by the back door,' said Dr Sue Mayer, GeneWatch's Director and author of the report. 'Industry and government hope to get around public concerns by using GM technology on non-food crops. We know it is difficult to contain GM crops inside a field or farm but GM grasses and trees will not even stay inside a country. Although people aren't going to eat them, the GM contamination threat to other plants remains'.

■ The full text of the report can be downloaded as a pdf file (240kb) from: www.genewatch.org/CropsAndFood/Reports/non-food_crops_part2.pdf

■ The above information is from GeneWatch UK's website which can be found at www.genewatch.org

GM products 'slip into foods'

By Sean Poulter

Millions of Britons are eating genetically modified ingredients every day without realising it.

According to research, many products and stores which claim to be GM-free are not.

After a consumer backlash, British manufacturers and supermarkets have gone to great lengths to try to ensure raw materials are GM-free.

But critics seized on the new findings as evidence of how difficult it is to hold back the GM tide once crops are allowed into the fields.

Scientists found that a Cow & Gate's 'GM-free' baby food – a dried vegetable casserole – contained low levels of genetically modified soya.

Three products sold under the popular Protoveg label as a meat substitute contained GM ingredients despite claims to be 'made from non-GM soya', researchers found.

The products, made by British firm Haldane Foods, are widely available at supermarkets and health food stores.

Similar products and a breakfast soya bran from health food chain Holland & Barrett also contained trace levels of GM ingredients.

A box of Farley's gluten-free reduced sugar rusks contained trace levels of the same material.

Premier Original Biscuit Cakes from Itona Products of Wigan were found to include GM traces despite claiming to be free of these ingredients.

But the use of impure seed overseas, particularly in the United States and Canada, cross-pollination in the fields and contamination of GM-free ingredients during shipping all create problems.

The GM contamination was discovered by the Food Safety Authority of Ireland.

EU and British laws allow manufacturers to have GM contamination of up to one per cent of contents without declaring the fact on the label.

It is only those companies which claim their foods are GM-free – when they are not – that may be guilty of breaking labelling laws.

Critics seized on the new findings as evidence of how difficult it is to hold back the GM tide once crops are allowed into the fields

But there is no evidence any firms identified were trying to mislead consumers.

Friends of the Earth's Pete Riley said: 'The companies involved need to do much more to tighten the quality control.

'Most rely on wholesalers who certify that their raw materials are GM-free. But it seems they should be carrying out their own checks.'

All the companies insisted the presence of GM material was accidental and outside their control. They stressed the levels were low.

Cow & Gate's parent, Nutricia, said: 'We do not use any GM ingredients and we are very careful about that.

'Where we use soya for protein content, we make sure it is from crops that are segregated and we DNA test it. This was not deliberate.'

Heinz, parent company of Farley's, says it operates a GM-free policy with ingredients and works hard to ensure none gets through.

Holland & Barrett said it regretted the discovery of the GM material and it had taken 'all practical steps' to avoid such a problem.

Haldane Foods said it went to great lengths to procure non-GM ingredients. Itona Products could not be contacted.

The GM positive foods

- Cow & Gate Dried Vegetable Casserole – GM soya
- Farley's Gluten-free Reduced Sugar rusks – GM soya
- Itona/Granny Ann Premier Original Biscuit Cakes – GM soya
- Protoveg Menu Sosmix with country herbs – GM soya
- Protoveg Menu Sosmix – GM soya
- Protoveg Menu Burgamix – GM soya
- Holland & Barrett Unflavoured Soya Mince Protein – GM soya
- Holland & Barrett Savoury Flavoured Soya Chunks – Unidentified traces
- Holland & Barrett Maize Meal – GM maize
- Holland & Barrett Soya Bran – GM soya

■ This article first appeared in *The Daily Mail*, May 2003.

Genetic modification of plants and food crops

Information from www.gmissues.org

How is a plant genetically modified?

Basically, genetic modification of a plant using the techniques of modern biotechnology involves:

- cutting out, using chemical 'scissors', one or more genes from the DNA of a plant, and inserting it into the DNA of another plant
- growing the plant and checking that it contains the inserted gene (i.e. that it is GM)
- checking that the inserted genes work as expected in the GM plant
- checking that the inserted gene is inherited unchanged in the GM plant's offspring.

Genes that are to be used to genetically modify a plant must be linked to pieces of DNA that control how they work. Genes (which encode proteins) need other pieces of DNA (known as promoters) to switch them on so that they will start working (expression).

Controlling gene expression

Some promoters cause the genes to which they are linked to be expressed all the time, whereas others allow expression only at certain stages of plant growth (e.g. flowering), or in certain plant tissues (e.g. roots or shoots). Some promoters are weak, whereas others are strong, and this determines how much of the gene product is made. By controlling gene expression, we can direct the plant's energy resources into growing or synthesising valuable molecules like starch or pharmaceuticals, as we wish. At present, in GM Bt corn or Bt cotton, the Bt gene (which makes the plant insect resistant) is switched on all the time, making it more likely that insect pests resistant to Bt will arise. There are therefore efforts to link expression of the Bt gene to a promoter that only switches the gene on to make insecticide when the plant is wounded, by being eaten by an insect, for example.

John Innes Centre

Getting genes into plants

There are two key methods that can be used to insert genes into plants:

I) *Agrobacterium*. A soil bacterium, *Agrobacterium*, has a natural ability to transfer its own genes into plants: it has been called 'Nature's own genetic engineer'. This ability can be harnessed by scientists to insert other genes of interest. Until recently, this method did not work well with cereals like maize or rice, and other methods had to be developed.

II) *Gene-gun*. The 'gene-gun' or 'biolistics' method can be used with all plant species, but is mainly used with cereals. Tiny gold or tungsten particles, coated with DNA, are fired into plant tissue.

Growing and testing GM plants

Plant cells or tissue into which genes have been inserted can be regrown in the laboratory by the use of plant hormones, and careful culture, into whole plants. The whole plants can then be tested for environmental (and food) safety.

Why genetically modify plants?

The use of GM in plant breeding aims to:

- increase crop yields beyond the maximum for existing varieties
- reduce post-harvest losses
- make crops more tolerant of stresses (cold, drought, salt, heat)
- make crops that do not exhaust soil fertility (make better use of nitrogen, phosphorus etc.)
- improve nutritional value of foods
- reduce reliance on chemical pesticides by producing pest-resistant crops
- develop alternative resources for industry such as starches, fuels, and pharmaceuticals.

Some of these aims involve transferring genes across species in a way that cannot be done by plant breeding. Whether it is SAFE to do this depends on which genes are being transferred, and this is addressed by safety assessments and regulations Whether it is ETHICAL to do so raises a different set of questions.

How is GM different from ordinary plant breeding?

Because we can identify which genes code for particular characteristics, and move these genes from one organism to another, we can produce a desirable combination of genes more quickly and easily by genetic modification than by breeding. Sometimes genes are moved between closely related organisms – for example, moving a gene from a weed that is naturally resistant to insects to a closely related crop to make the crop pest-resistant. Alternatively, genes can be moved between very different organisms, e.g. production of hepatitis vaccine in plants. In conventional breeding, it is impossible to move just one or two genes. Usually, whole chromosomes, containing thousands of unknown genes, are transferred.

Is plant breeding 'natural', or safe?

Many people are concerned that GM isn't natural and believe that conventional breeding is better because it follows the principles of natural selection, or uses natural mutations. However, it is just as possible, if not more so, to produce undesirable combinations of genes by conventional breeding: potatoes with dangerous levels of toxic glycoalkaloids and celery with high levels of chemical irritants have been produced by conventional breeding.

With cross-breeding, it is difficult to transfer genes between unrelated plants. However, plant breeders have devised ingenious techniques e.g. 'embryo rescue' to force crosses between species that wouldn't normally interbreed.

Plant breeders have for years used techniques to generate more variation than Nature produces: bombardment of a plant with mutagenic chemicals or radiation causes mutations randomly throughout its genes. From the plants that survive this treatment, plant breeders select the mutations they want. So long as the plant grows well and shows no toxicity, other mutations (maybe tens, maybe hundreds) will remain unidentified and uninvestigated.

Points to consider:

Why is it that the use of chemicals and radiation to produce random numbers and kinds of mutations as part of conventional plant breeding is acceptable, but changes to one or a few known genes by GM is not?

If it is just as possible to produce new unknown combinations of genes (possibly with higher levels of toxic substances; or unknown environmental effects) by 'conventional' breeding, shouldn't ALL new plants be tested in the same way as GM plants?

GM and farming

Even before our ancestors settled down to become farmers, humans were busy changing their environment: as bands of roving hunter-gatherers, they killed off many of the big mammals. Farming was an attempt to ensure a more secure food supply, and the aim has always been to prevent weeds, pests and diseases from competing with us for our crops. In medieval times in Europe, the farmer expected in a good year to be able to consume only one-third of his crop: one-third was saved for planting the following year and one-third was lost to pests. With the development of higher yielding crops through thousands of years of artificial selection, and intensive agriculture, productivity is higher although, in the developing world, up to 50% of crops are still lost to pests and diseases.

The past 50 years have arguably seen greater changes in agriculture than the previous 1000 years: high intensity agriculture is more efficient and productive, reducing prices and the risk of total crop failures like the potato famine of the 1850s, and consequent hardship and economic collapse. These agricultural changes all aim to achieve the goal of the first farmers: growing crops that only we can eat. Undoubtedly this has had an impact on the wildlife that depends on the same fields to exist – the weeds, the insects, birds and mammals – and has made major changes to the environment.

Considering the need to feed the rapidly growing world population and recognising that all agriculture is 'unnatural' and changes the environment, everyone agrees that future farming must be 'sustainable', but has different visions of what sustainable is. Three basic models can be thought of: high-intensity (high-input) agriculture; organic agriculture; and GM crop-based agriculture.

High intensity:

- High input (i.e. reliant on agrochemicals – fertilisers, herbicides and pesticides), and mechanisation to maintain high yield.
- For many crops, multiple crops can be grown each year.
- Requires new varieties to remain competitive; but many crops are reaching their biological and physical limits to yield. Cross-breeding slow to produced improved varieties; difficult to introduce new characteristics by crossbreeding within limited gene pool.
- Pesticide use can eventually lead to pest resistance, necessitating higher doses and new pesticides.
- High-intensity agriculture has a record of producing more than sufficient food to feed the world population, but at high environmental cost.

Organic farming:

- Attempts to minimise the impact on the environment and produce food by reducing or eliminating the input of agrochemicals such as pesticides, herbicides and 'chemical' fertilisers. Organic farming in the UK still only comprises around 1.5% of the total, but is growing at 40% per year, because of consumer demand for 'natural food'.
- Organic agriculture permits the use of 'natural' pesticides: based on plant products (e.g. Derris powder); or Bordeaux mixture (copper-containing sprays) and sulfur dusts to control fungal disease (obtained from natural sulfur deposits, the latter sometimes contain lead and other toxic heavy metal). Not all organic pesticides are specific for pests, but can also be toxic to non-target species of microbes, plants and animals.
- Uses 'traditional' crops and varieties produced by conventional breeding, but does not permit use of GM crops. Sprays made from a soil bacterium (*Bacillus thuringiensis*, 'Bt') that produce a toxic protein are used to control insect pests and are deemed 'organic', but GM plants that produce the protein are not.
- Reports of organic productivity highly variable – some claim higher productivity than high-intensity agriculture, but overall appears to be lower. Low yield agriculture requires more land. Unlikely it could feed current population of 6 billion, or future

population growth. Global scale-up of organic agriculture could generate many problems seen with current agricultural practice (e.g. greater susceptibility to pest or disease epidemics and crop failure). Historically, farming was all 'organic'.

GM crops:

- GM could enhance agricultural productivity by introduction of genes from same or other species (very slow, or not possible by crossbreeding) barriers to: break existing variety yield barriers; reduce reliance on agrochemicals by producing disease-, pest-resistant varieties; create optimally adapted varieties that do not exhaust soil fertility (i.e. make better use of nitrogen, phosphorus); decrease water requirement by adaptation to drought; reduce post-harvest losses to pests and improve nutritional value of foods; make use of marginal environments through adaptation to salt, cold or heat; develop alternative resources for industry such as fuels, starches, pharmaceuticals.
- As with any other pesticide, pest resistance to GM crops is ultimately likely, and indiscriminate use of pest-resistant plants will hasten their appearance.
- The use of plants as 'factories' to produce renewable, 'clean' resources for industry may help to replace environmentally damaging chemical industries and exhausted fossil-fuel supplies, but will inevitably require land for growing the industrial crops. Developing plants able to use marginal lands (too dry, salty or high in aluminium) and currently useless to agriculture, will result in wildlife habitats being lost.
- While pest- and herbicide-resistant plants are available, other developments important for securing a food supply (e.g. nitrogen fixation and stress-resistance) are way behind: achieving some of these goals is likely to take another 5-20 years. Field performance, as with any other variety, is likely to vary depending on environmental conditions and farming practice. GM crops are likely to help secure food supplies for future population growth, but the extent to which it will help is unknown.

What is sustainable?

- Today's high intensity agriculture is not sustainable long-term: in industrialised countries, 10 calories of fossil fuel energy are spent (mechanisation, production of agrochemicals etc.) to produce 1 calorie of food; a century ago during the agricultural revolution, the ratio was 1 calorie per 1 calorie food; and in hunter-gatherer society, it was 0.1 to 1.
- Yesterday's methods of agriculture are not adequately productive to feed today's population, let alone tomorrow's.
- Food production has so far kept pace with population growth with higher yielding crops and agrochemicals, but has incurred environmental damage.
- Economic sustainability in a competitive global market requires a competitive and efficient agriculture.
- Most crops require a lot of water – increasingly, in the next century, unpolluted fresh water will become a limiting resource across the world.
- All farming has some impact on the environment (like most activities to make life more secure for ourselves, farming isn't 'natural'). The best sustainable solution will probably require a combination of methods. Integrated pest management (IPM) uses biocontrols and rotates crops, creates refuges and uses agrochemicals in moderate amounts.
- GM is seen by some as incompatible with organic farming, but a combined GM-organic approach could have environmental benefits in reducing the use of artificial chemicals: for example, GM plants resistant to fungal disease could reduce the use of fungicides. Nematodes destroy nearly £70 billion in crops worldwide annually, and there are few options available for large-scale agriculture for controlling them: crop rotation is only partially successful at best, and crop protection relies on some of the most toxic and environmentally damaging pesticides in widespread use. Classical plant breeding has so far failed to produce effective nematode-resistant varieties, but GM varieties are in development.

Points to consider:

Many foods and other products are marketed as 'natural'. What does this really mean? What is the difference between a 'natural' chemical and any other?

The earth's resources are finite – how can it support an ever-growing human population?

Minor insect- or parasite-damage to food grown without pesticides is certainly 'natural', and generally not harmful to people. However, damage allows the growth of fungi, some of which produce mycotoxins – highly toxic 'natural' substances, some of which can cause cancer (e.g. aflatoxins in peanuts).

- The above information is from the website www.gmissues.org

What are the potential benefits?

Information from the Biotechnology and Biological Sciences Research Council (BBSRC)

For the environment

Crops can be modified to resist attack by specific pests. This both reduces the amount of pesticide that has to be used and leads to more targeted control, which protects beneficial species such as pollinators and the natural predators of pests.

Crops can be modified to tolerate a broad-spectrum herbicide (weedkiller) that is rapidly inactivated in the soil. This means that farmers need only apply this one herbicide instead of having to spray with several different ones. There is no longer any need to spray to prevent the emergence of weeds; weeds can be killed at any time during crop growth; and no residual soil-acting herbicides are needed. Better weed control reduces tillage, benefiting soil biodiversity, reducing erosion and helping to conserve moisture.

GM crops have the potential to increase crop yields, for example by reducing damage from pests and diseases, and by improving uptake of nutrients from the soil. This means that more food can be grown on the same land, leaving more land for wildlife conservation and other purposes.

GM crops can also help reduce energy consumption associated with the manufacture of agrochemicals. For example, US industrial data indicate that saving 5 million pounds (in weight) of formulated insecticides by growing crops genetically modified to resist attack by insects could save: 1,500 barrels of oil; 2.5 million pounds of waste; 150 thousand gallons of fuel; and 180 thousand containers and packages.

GM crops may be developed for non-food uses, for example for the production of valuable oils and starches that are currently made chemically. This offers a renewable and environmentally more friendly alternative to chemical manufacturing processes.

For the consumer

The first food from a GM crop to be sold in the UK was a tomato purée made from tomatoes modified to remain firm during ripening. The processing of the GM tomatoes produced less waste and used less energy than the processing of conventional tomatoes. The purée was cheaper than its conventional counterpart.

In future, genetic modification could be used to develop foods with added nutritional value, e.g. higher levels of vitamins, and foods with longer shelf life, and better texture and flavour.

Industrial data from an insect-resistant GM crop of cotton in the USA showed that 70% of the crop required no insecticide spray, and the remainder was sprayed only once instead of 4-6 times.

Between 1996 and 1998, total insecticide applications to GM cotton were reduced by over 3 million litres compared to non-GM cotton.

Preliminary trials of herbicide-tolerant sugar beet in the UK show that effective weed control can be achieved with two applications of the herbicide, compared with up to seven or eight sprays of selective herbicides. Spraying costs were £24 per acre for the GM crop compared with up to £140 per acre for the non-GM crop. The weeds growing in the GM crop acted as a decoy for pests, and when sprayed they continued to provide a habitat for insects that was missing in the conventionally sprayed crop, so biodiversity was enhanced in the GM crop.

For the farmer

Genetic modification offers new ways of controlling pests and diseases. For example, naturally occurring genes that confer resistance to attack by fungi could be introduced into commercial varieties. The microbial genes that code for compounds that specifically kill different pests are being introduced into a range of crops including maize.

Genetic modification offers new ways of improving crop yield: for example, oilseed rape may be modified to reduce seed loss due to pod shatter. The technology could also improve product quality: for example, wheat might be modified to enhance its bread-making properties

■ The above information is from *GM agriculture in the UK?* published in 1999 by the Biotechnology and Biological Sciences Research Council's (BBSRC). For more information visit their website: www.bbsrc.ac.uk

Myths & GM

Information from Monsanto

Re[illegible]tion of Alex[illegible]ocqueville (1885-[illegible]at 'The public will always belie[illegible] a simple lie rather than a complex truth,' perhaps it is time to develop a mantra of maxims about biotechnology to counteract the simple lies. Wherever your biotech information comes from, is it trustworthy and true? Given that GM is primarily a science, and not an art, just how sound is the science behind the information you receive? In short, is it peer-reviewed?

Peer review is accepted globally as the best defence against basic mistakes and very occasional attempts by some to defraud the scientific system with totally misleading conclusions. Let us examine one example of anti-biotech claims from 1999 which still exists today.

'GM corn kills monarch butterflies'

When monarch butterflies became the subject of concern once the initial results of a laboratory study, conducted by researchers at Cornell University in the US received publicity in 1999, www.nature.com/cgi-taf/DynaPage.taf?file=/nature/journal/v399/n6733/abs/399214a0_fs.html understandably many become emotionally attached to these beautiful creatures. Reputable scientists had discovered that large amounts of Bt corn pollen fed to monarch larvae affected the butterfly's survival. Bt corn produces a protein that protects it against a pest called the European Corn Borer which eats the stem of the corn rendering it useless; the protein kills the pest before it can harm the crop, thereby reducing the use of insecticides by farmers who also benefit from increased harvest yields. www.monsanto.com/monsanto/content/our_pledge/transparency/prod_safety/yieldgard_corn/pss.pdf

Before that research was completed, there were other reports which caught the media's attention. Indeed, one website called 'Digest of Reports about the butterfly-killing corn' quoted an independent non-profit alliance of scientists called the Union of Concerned Scientists who said they were dismayed but not surprised that risks had not been properly studied. www.organicconsumers.org/ge/butterfly.cfm It also quoted from the BBC news.bbc.co.uk/hi/english/sci/tech/newsid_347000/347638.stm

In August 2000, Greenpeace reported that responsibility lay with Kellogg's in America to stop using 'butterfly-killing corn' as field evidence had confirmed that 'gene-altered corn kills monarch butterflies.' www.greenpeace.org/pressreleases/geneng/2000aug21.html Meanwhile, Friends of the Earth said 'Research is needed in this country to find out whether UK butterflies and beneficial insects will be threatened by GM crops . . . It may be that GM crops will be just as harmful to beneficial insects and biodiversity as the pesticides used at the moment'. www.foe.co.uk/pubsinfo/briefings/html/20000208165738.html see 'Wildlife Fears'. And Jeremy Rifkin, a long-time biotechnology industry critic, said he saw the Cornell research as the 'Silent Spring' of genetically modified foods. 'In the next 24 months, we're going to see an industry go down," Rifkin said. 'There are more and more questions being raised.' (*Wisconsin State Journal*, 28 June 1999)

As a result of the Cornell study, comprehensive research was initiated in 1999 by scientists in the US and Canada. As these studies progressed to their conclusions, researchers were open with their findings and presented results from laboratory and field studies at scientific symposia and workshops. Six peer-reviewed studies were published www.pnas.org/cgi/content/full/98/21/11937 in a technical journal called the *Proceedings of the National Academy of Sciences* (PNAS) in September 2001. The transparency of these studies including the unusual willingness of scientists to discuss their findings, was crucial to the process that helped everyone understand whether there was any real risk to monarchs or not.

At the end of the field studies process, two eminent scientists who are both specialists in this area (one of them from Cornell University where the original laboratory tests were conducted) summed up their work in a letter sent to the US Environmental Protection Agency (EPA). The letter states: 'In the more than two years that this research has been under way, the research community and other stakeholders have presented, discussed, and weighed the information in multiple public forums, including workshops and symposia at scientific meetings. These open discussions have ensured the data generated have been available for review and comment consistently from the beginning of this process.'

Bt corn and monarch butterflies were once again given the 'all clear' in 2002 by the United States Department of Agriculture, see www.ars.usda.gov/is/AR/archive/feb02/corn0202.htm

There is always an answer to every allegation and all of the claims in 'Seeds of Doubt' for example, are answered in the document 'Correcting the Myths' at www.asa-europe.org/pdf/longversion.pdf The difficulty is often to know where to look for answers. For example, in 2003, of the 7 million farmers who chose to plant 167 million acres of GM crops in 18 countries, three-quarters of them were resource-poor growers in developing countries. Such facts are available from www.isaaa.org/kc/CBTNews/press_release/briefs30/es_b30.pdf

■ If you would like more information on recent GM myths, please ask your question at www.monsanto.co.uk/comments/comments.html

GM – update and suggestions for action

Information from the Soil Association

Current situation

Those opposing GM crops have won a significant victory in the battle to protect GM-free food and farming. Despite the UK government giving the green light for the commercial planting of GM maize for animal feed, this GM crop will not be grown in the UK.

We are extremely disappointed that the government chose to disregard public opinion, ignoring the evidence that clearly shows GM crops should not be grown in the UK,[1] and that the public do not want GM. The good news is that Bayer, who own Chardon LL maize, has abandoned plans to get its product approved.

Bayer withdraws GM maize application

On the last day of March, Bayer withdrew its application, blaming tough government restrictions. We believe the truth to be that they were caught out by their own hype. GM companies have claimed that GM crops need fewer chemical sprays. Bayer's GM maize was grown during the government's field scale trials, with use of one weed-killing spray. But Soil Association research in the USA and Canada had already shown GM maize grown commercially needed at least two weed-killers.

Unfortunately for Bayer, the British government took them at their word, and said that their GM maize could only be grown using one weed-killer. Based on experience in America,[2] Bayer knows that won't work in practice. We believe that it is this that has led Bayer to withdraw the GM maize.

Peter Melchett, the Soil Association's Policy Director, said, 'This is wonderful news . . . It's a victory for reason, common sense and the will of the general public.'

What the government decision means:

Beet and oilseed rape opposed: The UK government has opposed EU approval for the commercial cultivation of GM beet and oilseed rape as grown in the GM crop trials.

GM industry must pay for damage: According to the government's GM committee, the GM industry will be liable for any damage their GM crops cause.

Take action

Ask your supermarket if their own-brand non-organic milk is GM-free.

The supermarkets, major food companies and consumers all say that they will not buy GM food. The only remaining outlet for GM maize is as feed for livestock, particularly dairy cows. Many non-organic livestock are currently being fed imported GM soya and maize. From April 2004, GM animal feed must be labelled, so dairy farmers can easily exclude it from their animals' diet. But the non-organic consumer cannot choose to exclude GM from their diet because the milk from cows fed on GM feed is not labelled as such.

Only a GM-free policy from supermarkets and dairy producers will ensure milk is GM free. This is the best way of stopping GM maize being grown. If you don't want to drink milk or eat meat from animals fed on GM crops, let your supermarket know.

Ask the supermarket what they are going to do to make sure GM feed is not used. Please send us copies of any written replies you get.

ASDA 0500 100 055
Co-op (CWS) 0800 317 827
Marks & Spencer 0207 268 1234
Safeway 01622 712 987
Sainsbury's 0800 636 262
Somerfield 0117 935 6669
Tesco 0800 50 55 55
Waitrose 0800 188 884

Further points to tell the supermarkets:

- As an organic consumer you are concerned that GM maize could contaminate non-GM seed stocks and non-GM maize in the field.
- Organic farmers growing sweetcorn for retailers, vegetable box schemes and anyone growing sweetcorn in allotments and gardens will also be affected.
- GMOs go against the principles of organic production.
- Companies are now moving to ensure that no GM animal feed is used for their meat and dairy products. For example, Marks and Spencer, Co-op, Grampian chickens and Kentucky Fried Chicken.
- Non-GM feed policies will be easier to implement due to the requirement for GM feed to be labelled.

How to support the work of the Soil Association

The Soil Association is a membership charity, we urgently need your support to continue our work. As public support for the Soil Association continues to grow, our ability to influence the thinking and policies of government and big business grows with it. Join us today and help us to continue campaigning for sustainable agriculture and organic food. You can join the Soil Association on our website, over the phone or by writing to us.

Footnotes

1 The Independent Science Panel, *The Case for a GM-Free Sustainable World*, 15 June 2003, The full report is available at the ISP's website: http://www.indsp.org/

2 Benbrook CM (2003) *Impacts of Genetically Engineered Crops on Pesticide Use in the United States: The First Eight Years*, BioTech InfoNet, Technical Paper No 6, Nov 2003. The report analyses agrochemical (herbicide and insecticide) use over eight years (1996-2003) on the three main GM herbicide tolerant (HT) crops in the US (soya, maize and cotton) and the two main GM Bt crops (maize and cotton). The report can be downloaded at http://www.biotech-info.net/technicalpaper6.html

■ The above information is from the Soil Association. For further information visit their website which can be found at www.soilassociation.org

GM *crops – going against the grain*

Nearly 800 million people go hungry every day because they cannot grow or buy enough food. One in seven children born in the countries where hunger is most common die before they are five years old.

Many governments, companies and institutions are promoting genetically modified (GM) crops as a response. It is claimed GM technologies will increase food production, reduce environmental degradation, provide more nutritious foods and promote sustainable agriculture. But can GM crops really alleviate world hunger?

ActionAid believes that food security can only be achieved by addressing poverty, matching technologies to local needs, promoting basic rights, protecting biodiversity, and supporting informed choice and participation for poor people. This report – which is based on evidence from Asia, Africa and Latin America – concludes that GM crops are unlikely to contribute to any of these objectives. The expansion of GM is more likely to benefit rich corporations than poor people.

Key statistics

- GM crops covered 58 million hectares worldwide in 2002 – an area two and a half times the size of the UK.
- Only 1% of GM research is aimed at crops used by poor farmers in poor countries.

- It can cost up to $300 million to develop a GM crop and the process can take up to 12 years.
- A small range of GM crops that might address poorer farmers' needs are being researched but they stand only a one in 250 chance of making it into farmers' fields.
- The four corporations that control most of the GM seed market had a combined turnover from agrochemicals and seeds of $21.6 billion in 2001.
- 91% of all GM crops grown worldwide in 2001 were from Monsanto seeds.

Can GM crops help eradicate poverty?

It is not the interests of poor farmers but the profits of the agrochemical industry that have been the driving force behind the emergence of GM agriculture. Four multinational corporations – Monsanto, Syngenta,

Bayer CropScience and DuPont – now control most of the GM seed market. Some 91% of all GM crops grown worldwide in 2001 were from Monsanto seeds. By linking their chemicals to seeds via GM technologies, these corporations have been able to extend markets for their herbicides and pesticides.

GM crops are unlikely to help eradicate poverty because yields seem to be no more than non-GM crops and sometimes need more chemicals. Yields from GM soybeans are no higher than those from high-yield conventional varieties. In one study, Monsanto's GM soya had 6% lower yields than non-GM soya and 11% less than high-yielding non-GM soya.

Insecticide use on GM cotton has fallen in some locations, but these gains may be short-lived as insects develop resistance to the insecticide that the cotton expresses. In time, farmers may need to invest in more, not fewer, chemicals. This also applies to chemical use on herbicide-resistant GM crops, which has gone up rather than down as farmers use chemicals more frequently and/or in greater amounts. Herbicide use per hectare in Argentina has more than doubled on GM fields compared to conventional varieties.

GM crops are ineffective in tackling the underlying political and economic causes of food insecurity: poverty and inequality. The new GM technologies do not address the essential constraints facing poor farmers including lack of access to land, water, energy, affordable credit, agricultural training, local markets, decent roads, grain stores and infrastructure. In fact, GM could be disastrous for small-scale farmers as the costs are much higher and they risk falling into debt.

Do GM crops meet the needs of poor farmers?

GM varieties do not meet the needs of poor farmers who rely on affordable, readily-available supplies of seeds for a range of crops to meet diverse environmental, consumption and production needs. Poor communities need investment in low-cost, low-input farmer-friendly technologies, building on farmers' knowledge. GM seeds, by contrast, are targeted at large-scale commercial farmers growing cash crops in monocultures. GM crops could undermine food security by wasting the scarce resources of poorer farmers and developing countries.

Most research and development in GM agriculture is conducted by the private sector. Less than 1% of all GM research is directed at poor farmers.

GM research in Africa, for instance, focuses on export crops such as cut flowers, fruit, vegetables, cotton and tobacco, which are grown in large-scale commercial plantations in Kenya, South Africa and Zimbabwe. In Kenya, only one out of 136 intellectual property applications for plants were for a food crop; more than half were for roses.

Do GM crops threaten basic rights?

Farmers in developing countries have evolved complex, cheap and effective systems to save, exchange and use seeds from one harvest to the next. Patented GM seeds threaten to erode these rights and practices, to displace or contaminate seed supplies, and to increase farmers' dependence on private monopolised agricultural resources.

Up to 1.4 billion people, including up to 90% of farmers in Africa, many of them women, depend on saved seed. Yet the proliferation of intellectual property regimes that comes with GM seeds threatens centuries-old practices of saving and exchanging seeds.

GM seeds must usually be bought each season. Before they can obtain and use the seeds, farmers have to sign a contract with the company obliging them to pay a royalty or technology fee, to agree not to save or replant seeds from the harvest, to use only company chemicals on them and to give the corporation access to their property to verify compliance.

Having to buy external supplies of seeds and pesticides leaves farmers more economically and agriculturally dependent on corporations. The technology fee makes such seeds prohibitive for the poorest farmers who lack access to credit. The contracts are complex and easily misunderstood by farmers, especially those who are illiterate.

The biotech industry continues to develop a set of GM crop technologies – Genetic Use Restriction Technologies (GURTs), which have been dubbed 'terminator technologies' – that produce sterile seeds: if saved and planted from one year to the next, they would have no yields at all.

Area and type of GM crops grown in developing countries, 2002

	Area (millions of hectares)	% of developing country total	Type of GM crop
Argentina	13.5	85.0	Herbicide tolerant (HT) soybean, HT maize
China	2.1	13.0	Bt cotton
South Africa	0.3	1.9	Bt maize, HT cotton, HT soybean
India	<0.1	<1.0	Bt cotton
Uruguay	<0.1	<1.0	HT soybean
Mexico	<0.1	<1.0	Bt cotton, HT soybean
Indonesia	<0.1	<1.0	Bt cotton
Colombia	<0.1	<1.0	Bt cotton
Honduras	<0.1	<1.0	Bt corn
Total	15.9	100.0	

Source: James C, ISAAA 2002

Do GM crops threaten biodiversity?

GM crops threaten to reduce the agricultural and crop diversity that is the basis of poor farmer livelihoods and developing country food sovereignty. Three-quarters of the original varieties of agricultural crops have been lost from farmers' fields since 1900 as industrial and export-led agriculture has encouraged the widespread monoculture cultivation of a few crop varieties for a more uniform global market. GM crops threaten to erode biodiversity still further.

In addition, GM crops pose known threats to other plants and insects. They can cross-pollinate with non-GM plants, endangering diverse original varieties, particularly in developing countries. They are likely to require bigger and more frequent doses of pesticide as weeds and insects develop resistance to chemicals. They may threaten beneficial insects and thus disrupt natural pest management systems. GM crops engineered to produce pharmaceutical drugs could easily end up in local food supplies.

Biosafety regulations could address some of these problems and threats to biodiversity, but many countries do not have them, or the capacity to develop them. In Zambia, just one person, who has no previous experience of developing national policy or prior knowledge of the issues, is responsible for drafting national biosafety policy.

Nor is regulation enough where national capacity to evaluate and monitor risks is weak. In Brazil, a ban on the commercial cultivation of GM crops did not stop GM soya seeds being smuggled in from Argentina and planted across huge areas. In Pakistan, ActionAid has investigated the impact of illegally-planted GM cotton. Hundreds of farmers who bought the so-called 'miracle' seed on the black market in the hope it would increase their harvests lost around 70% of their crops.

Do GM crops enhance informed choice and participation for poor people?

Developing country governments are under huge pressure to accept GM crops, put scarce public resources into GM research and open their doors to biotech corporations before their people have been properly informed, consulted and agreed to accept, or reject, GM. Poorer farmers and communities are being sidelined in debates and decisions about GM technology.

The widespread adoption of GM crops seems likely to exacerbate the underlying causes of food insecurity

In South Africa, for example, GM crops have been planted without prior public consultation or involvement in decision-making and without environmental studies on their impact.

Even if GM research takes place in the public sector it may not address the needs of poor farmers because most genes and processes are now patented by corporations. In partnerships between public research organisations and corporations, control and decision-making tend to remain firmly in the hands of corporations who acknowledge that their goal is to create new markets and improve their public image.

If poorer people were more involved in setting agricultural research agendas, they would probably opt not for GM crops, but for other agricultural solutions.

Conclusion

The widespread adoption of GM crops seems likely to exacerbate the underlying causes of food insecurity, leading to more hungry people, not fewer. To have a lasting impact on poverty, ActionAid believes policy makers must address the real constraints facing poor communities – lack of access to land, credit, resources and markets – instead of focusing on risky technologies that have no track record in addressing hunger.

Recommendations

- Donors and governments should address the wider causes of food insecurity – land, credit, agricultural training and infrastructure – before putting resources into GM crops.
- They should introduce a moratorium on the further commercialisation of GM crops until more research has been carried out into the socio-economic, environmental and biodiversity impacts of GM crops, particularly in developing countries.
- Poorer farmers and communities should be enabled to participate more in national GM debates and policy-making.
- Genetic resources for food and agriculture should be exempt from intellectual property requirements.
- Farmers' rights to save and exchange seeds should be recognised under the intellectual property rules of the World Trade Organisation (WTO) and should be protected in developing country intellectual property rights legislation.
- Governments should introduce competition rules to prevent private sector monopolies and effective institutions to enforce them.
- The potential impact of GM crops on food security, poor farmers and biodiversity should guide the development and implementation of national biosafety frameworks.
- Funding for public sector agricultural research should be increased and should specialise in support for sustainable, farmer-led agriculture.

■ The above information is from the Executive summary of *GM crops – going against the grain*, produced by ActionAid. Visit their website at www.actionaid.org

GM losers

Why it isn't paying to plant genetically modified crops. Evidence from some of the countries that have been growing it . . .

Who's growing it?

We are told that British farmers will miss out if we do not join the countries already growing genetically modified crops. Just three countries (USA, Canada and Argentina) grow 98% of all GM crops and North America alone produces three-quarters of all GM output.

In 2001 one company, Monsanto, planted 91% of the total area devoted to commercial GM crops.

Wide-scale adoption of GM in North America might suggest that farmers are embracing the technology because it is more profitable. But the marketing of GM seed in North America and elsewhere is achieved through an aggressive promotion policy designed to rapidly capture market share and create an irreversible shift to GM seeds.

A 1998 study by Iowa State University revealed the uptake of GM crops was driven by farmers believing marketing claims. More than half the farmers planted herbicide-tolerant GM soya because they believed it would produce higher yields than conventional varieties. Analysis showed the opposite was true, with GM yields down by 5-10%.

A US Department of Agriculture report, released June 2002, concluded that 'perhaps the biggest issue raised by these results is how to explain the rapid adoption of GE crops when farm financial impacts appear to be mixed or even negative'.

Fortunately, beyond cotton and the massive animal feed sectors served by soya, oilseed rape and maize production, few producers seem to be buying into the technology. US farmers are largely not growing transgenic sugar beet, potatoes or sweetcorn, despite all of them having been approved for cultivation for some time. Packers and processors have not been accepting these crops. Their concern is that doing so might jeopardise their markets for products intended for direct human consumption.

What are their profits like?

GM seeds cost more than conventional seeds but GM products fetch a lower price. In addition to lower farm profitability, GM crops have been a market failure internationally. Since introducing GM crops North American access to the markets of Europe, Japan, Korea and New Zealand has been seriously restricted, resulting in billions of dollars worth of lost exports and a collapse in farm-gate prices.

The USDA farm survey of 350 Iowa farms in 2000 reported data on yields, and fertiliser, herbicide and seed costs. Analysis of these figures showed there had been no economic on-farm benefit of GM crops to counter massive falls in prices.

US maize prices are at their lowest for 30 years because there is no demand for GM maize, but production costs have not fallen. Since GM Bt corn was introduced, exports to the EU have fallen from millions of tonnes to almost zero. In 1996, before GM crops were introduced, US maize farmers made a profit of $1.4 billion. Last year they lost $12 billion.

The US share of the world soya market has also taken a tumble since they started growing GM. Canadian oilseed rape exports to the EU were worth $180 million in 1996 – they are now down to zero!

Lost export markets and falling farm-gate prices caused dramatic increases in US farm subsidies which were meant to have fallen over the last few years but instead rose by an estimated $3 to $5 billion annually, in parallel with the growth in GM acreage.

In total GM crops may have cost the US economy at least $12 billion from 1999 to 2001. The US farm sector has become highly unstable, with record levels of farm bankruptcies.

The cash-hit Argentinian Government spent US$200 million to help farmers switch from GM crops and recover their export markets. Monsanto captured 90% of the soy seed market in Argentina with a sow-now-pay-later scheme. However they have since had to write off $1.8 billion (one-third of the company's tangible net assets) when the economy of Argentina collapsed.

Contamination problems

Many non-GM farmers in the USA and Canada are finding it nearly impossible to grow GM-free crops. Seed stocks have become contaminated and there is a high risk of their crop being contaminated by their neighbours', even if they plant normal varieties.

GM contamination and the lack of segregation have caused major disruption at all levels of the industry – seed resources, crop production, food processing and bulk commodity trading. It has undermined the viability of the North American farming industry and made the whole food processing and distribution system vulnerable to costly and disruptive contamination incidents.

In 2001 traces of GM potato were found in snacks exported to Japan. Japanese importers instituted strict testing protocols and the US lost 37% of its huge Japanese potato market. In response the US Potato Board has had to institute a costly programme to remove GM potatoes entirely. Monsanto closed its potato division in 2001.

In September 2000, just 1% of an unapproved GM maize called 'Starlink' contaminated almost half the national US maize supply. Designed for animal feed, the product was not given a licence for human consumption because it contained elements that trigger food allergies. Processors all over the world panicked. Japan immediately halved its imports of American maize and South Korea banned US maize altogether. Compensation cost the company, Aventis, one billion dollars.

Contamination has caused the loss of nearly the whole organic oilseed rape sector in the province of Saskatchewan, at a potential cost of millions of dollars. GM contamination has led to a proliferation of lawsuits and the emergence of complex legal issues.

Legal battles

One of the most unpleasant outcomes of the introduction of GM has been farmers being accused of infringing company patent rights. Growers are being prosecuted for not paying 'intellectual property rights' when their normal crops get polluted by neighbouring GM pollen. A Canadian farmer whose crop was contaminated by GM was successfully sued by Monsanto for $400,000.

While biotechnology companies are suing farmers, farmers themselves are turning to the courts for compensation from the companies for lost income and markets as a result of contamination. In Canada a class action has been launched on behalf of the whole organic sector in Saskatchewan for the loss of the organic oilseed rape market.

Consumers are also trying to fight back with legal measures. In Oregon they won the right to a ballot proposing the labelling of GM products. It was the first US state to take such a measure. The corporations responded with a propaganda campaign with a budget (including $1.5 million from Monsanto) forty times that of the consumers who, unsurprisingly, were defeated.

Yet, despite the saturation of arguments such money can buy, North American farmers are beginning to question seriously the development of GM crops.

In 2002 many US farm organisations urged farmers to plant non-GM crops. The US and Canadian National Farmers' Unions, American Corn Growers' Association, Canadian Wheat Board, organic farming groups and more than 200 other groups are lobbying for a ban or moratorium on the introduction of the next major GM food crop, GM wheat.

With the support of several farming organisations, federal legislation was tabled in Congress in May 2002 to introduce GM labelling and liability rules in the US. Nearly every country in the world now labels GM products apart from the countries that are growing it.

Which future do you want?

There is a battle being waged to capture world food markets with patented seeds and paired herbicides.

The marketing strategy seems to be . . . Promise everything. Spend big on public relations, farmer advertising and government lobbying. Give away seeds.

Between 1998 and 2000 Monsanto sold $7.5 billion worth of related Roundup herbicide. Meanwhile US farm-gate soya prices fell by more than $2 a bushel and soy farmers are losing billions.

AgBiotech alliances have the capacity to manipulate the agricultural sector and exert a powerful influence on governments. Ever wondered why the Blair government is so keen to champion GM?

The government minister with official responsibility for regulating biotech companies, Lord Sainsbury (who also happens to be Labour's biggest single donor), has made millions on GM food shares. His shares in Innotech rose from £26.9M in 1998 when he became Minister for Science and Innovation to £42.6M in 2002. While in office he has also overseen a massive 300% increase in public funding for the Sainsbury Laboratory which researches GM. If we allow the vested interests of our government and biotech lobbyists to commercialise GM crops in the UK, our farming industry could ultimately have to face export bans and undertake costly, if not impossible, clean-ups to protect markets.

Commercialisation of GM crops in the UK would be against consumer demand, would lose Britain its valuable status as a relatively safe haven from GM, and would render organic farming virtually impossible.

Much of the world is already awake to the danger and over fifty countries have placed restrictions on the growth or import of GMOs. Resistance to GM in the UK has already delayed commercial planting by at least four years and sent shockwaves through the industry. Let's not get on a sinking ship.

We can stop GM.

■ For more information on what you can do, contact: Totnes Genetix Group, Tel: 01803 840098 Website: www.togg.org.uk E-mail: info@togg.org.uk

Closing the stable door after the GM horse has bolted

How can we monitor the long-term health effects of GM foods once they're released? At a meeting held to answer this question Elisabeth Winkler asks: shouldn't we do some serious health testing before GM foods are let loose?

I am at the Food Standards Agency (FSA) at Aviation House in Holborn – its grand stone entrance fronts glass-pannelled offices. The FSA, a government agency, is part of the establishment (symbolised by the stone portal) but prides itself on being as transparent as its glass edifice.

In modern style, the FSA's Advisory Committee on Novel Foods and Processes is having an open meeting about its work. This 'non-statutory, independent body of scientific experts' is looking at how to monitor the long-term health effects of novel foods after they have been approved.

Novel foods includes GM foods and that's why the Soil Association – represented by myself and Gundula Azeez – is there. The audience also includes the usual suspects such as Monsanto, Syngenta and CropGen.

The divisions soon emerge. First we discuss, do we need GM? 'We in the west don't,' says one member of the audience stoutly. 'But the developing world does.' Exasperated snorts greet my argument that starvation has more to do with food distribution than GM. There was much discussion on the miracles that GM will deliver to the third world. To question this credo makes you look very heartless indeed. GM-feeds-the-world disciples, however, are on such high moral ground, they may be at risk from vertigo.

Another pro-GM voice says that the ordinary man in the street cannot possibly grasp the intricacies of GM, when Joe Public cannot even understand that eating too much leads to obesity. The same voice insists that the public, like himself, should rely on the genetic experts. I say that some people use other indicators to make food choices, such as 'gut instinct'. More snorts and derisive laughter follow, like MPs braying in parliament.

I feel like the Man who lit his Cigar before the Royal Toast for even mentioning this word. (Mind you, it is 'gut instinct' which makes animals choose non-GM feed over GM-feed every time, according to reports from US farmers.)

The British Dietetic Association ask if there is any post-market monitoring in the US. 'How many of us have recently visited the US and experienced short-term effects from eating food there?' replies one committee member ingenuously.

The sub-text is familiar ('they've been eating GM food in the US for years with no problems'), a spurious argument considering that GM food is unlabelled and health effects could be cumulative and long-term. (And, no, there is no post-market monitoring in the US.)

Gundula Azeez says that post-marketing monitoring is likely to only pick up gross problems. 'The committee must make its limits explicit otherwise the public will be lulled into thinking GM is safe. Also each GMO is unique, and individual constructs could lead to individual problems,' she says.

The committee reassures us that problems picked up would lead to further investigation. But surely health testing should be done before a novel food is approved?

The FSA funded the world's first and only human feeding trial on GM in 2002. A tiny sample of volunteers eating one GM meal resulted in human gut bacteria taking up GM genes. The panel's chair Professor Mike Gasson agrees that this result was an 'unexplained loose end' although he does not agree that gene transfer (when the GM gene escapes to another organism) had occurred.

If it was not gene transfer, then what was it? Surely this 'loose end' is worthy of a follow-up study? There have been only five independent studies in the world looking at the effects of animals eating GM – all five found negative results and yet none has been followed up. Why so many 'unexplained loose ends'? Excuse me, but I don't need gut instinct to tell me that is not very scientific.

■ Elisabeth Winkler is editor of the Soil Association's membership magazine *Living Earth.*

The gene revolution

Great potential for the poor, but no panacea

Biotechnology holds great promise for agriculture in developing countries said *The State of Food and Agriculture 2003-04*, the UN's Food and Agriculture Organisation in its annual report.

In the few developing countries where transgenic crops have been introduced, small farmers have gained economically and the use of toxic agro-chemicals has been reduced, FAO said.

'Transgenic crops have delivered large economic benefits to farmers in some areas of the world over the past seven years,' the report said. In several cases, per hectare gains have been large when compared with almost any other technological innovation introduced over the past few decades.

But so far only farmers in a few developing countries are reaping these benefits; basic food crops of the poor such as cassava, potato, rice and wheat receive little attention by scientists. 'Neither the private nor the public sector has invested significantly in new genetic technologies for the so-called "orphan crops" such as cowpea, millet, sorghum and tef that are critical for the food supply and livelihoods of the world's poorest people,' said FAO Director-General Dr Jacques Diouf.

'Other barriers that prevent the poor from accessing and fully benefiting from modern biotechnology include inadequate regulatory procedures, complex intellectual property issues, poorly functioning markets and seed delivery systems, and weak domestic plant breeding capacity,' he added.

Agriculture will have to sustain an additional 2 billion people over the next 30 years from an increasingly fragile natural resource base. The challenge is to develop technologies that combine several objectives – increase yields and reduce costs, protect the environment, address consumer concerns for food safety and quality, enhance rural livelihoods and food security, FAO said.

Agricultural research can lift people out of poverty, by boosting agricultural incomes and reducing food prices. More than 70 per cent of the world's poor still live in rural areas and depend on agriculture for their survival. Agricultural research – including biotechnology – holds an important key to meeting their needs.

Biotechnology should complement – not replace – conventional agricultural technologies; it can

- speed up conventional breeding programmes and may offer solutions where conventional methods fail
- provide disease-free planting materials and crops that resist pests and diseases, reducing use of chemicals that harm the environment and human health
- make available diagnostic tools and vaccines to help control devastating human and animal diseases
- improve the nutritional quality of staple foods such as rice and cassava and create new products for health and industrial uses.

But poor farmers can benefit from biotechnology products if they 'have access to them on profitable terms', the report said. 'Thus far, these conditions are only being met in a handful of developing countries.'

Sources:

The gene revolution: great potential for the poor, but no panacea. Food and Agriculture Organisation of the United Nations. http://www.fao.org/newsroom/en/news/2004/41714/index.html

The State of Food and Agriculture 2003-2004. Agricultural Biotechnology; Meeting the needs of the poor? Food and Agriculture Organisation of the United Nations, Rome 2004. http://www.fao.org/docrep/006/Y5160E/Y5160E00.HTM

■ The above information is from CropGen's website which can be found at www.cropgen.org

Against the grain

Consumers don't want to eat GM products, so researchers are looking for non-food ways to use the crops. But cottons, golf courses and plastics aren't safe either, warns Sue Mayer

The biotechnology industry needs to find other uses for its GM crops – uses which it hopes won't upset the public. Billions of dollars have been invested in developing crops, and intellectual property rights have been put in place that should allow the profits to roll in, but the resistance of people in Europe and many other parts of the world has upset the industrial dream of a GM future.

Prime targets for GM are the so-called 'non-food' uses: grasses, flowers, trees, cotton, and a range of different crops being modified to provide the raw materials for the industrial production of biofuels, oils, starches and plastics.

So, if you don't have to eat them, are there any real reasons to worry? In a word, plenty. Non-food uses are likely to bring in contamination of non-GM crops and nature by the back door. This much is clear if you consider what may be on the market soon.

Perhaps the most alarming development is GM herbicide-tolerant amenity grasses. Particularly in the US, there is a search for the perfect lawn – one which is low-maintenance, weed-free, uniform and that can survive stressful environments, such as prolonged periods of drought. Monsanto, in partnership with Scotts, a lawn and garden products company, is seeking to commercialise a GM herbicide-tolerant creeping bent grass in the US. The original application was withdrawn, but a decision on a new application, filed in 2003, is expected shortly. Experimental GM golf courses have already been planted.

The problem is that grasses are difficult to contain. They are freely wind-pollinating, perennial and often reproduce via underground shoots. Grasses spread internationally on wool, and in lawn and bird seed mixes, so attempts to isolate GM grasses will probably prove futile over time. Golf courses and gardens are often close to natural habitats and farmland.

No one, it seems, has considered the international implications of this development. Britain has worried about GM forage grasses for animals, but not amenity grasses. GM herbicide-tolerant grasses could pose weed problems for farmers and lawn-keepers alike, as well as having a very real potential to establish themselves as an alien invader.

Trees, while less advanced commercially, pose similar kinds of problems in terms of international contamination, and herbicide tolerance is another favourite of GM tree producers. It makes economic sense for the owners of the genes to use them as widely as possible, which is why Monsanto also has a toe in the GM tree water.

But there are also more familiar GM crops looking for a new role in life. The interest in biofuels to replace fossil fuels has led to the suggestion that GM herbicide-tolerant oilseed rape and sugar beet could both be used to improve production efficiency. This would open a new market for crops that have been rejected for food use. However, the contamination threat to non-GM food crops will be very real, especially with oilseed rape. And both oilseed rape and sugar beet have wild relatives in Britain with which they can hybridise.

Rather unsuccessful attempts have been made to turn oilseed rape and other oil crops into producers of specialist oils and plastics for industrial uses. The idea is that a particular oil produced by plants such as jojoba and coriander could be produced more efficiently in a domesticated crop. However, problems have arisen because producing the fatty acids that make up oil is much more complex than was once thought.

Fatty acids have at least three roles in plants – as a constituent of membranes, in cell signalling, and for energy storage. Unfortunately these are not controlled by separate pathways, and when novel fatty acid synthesis has been induced by GM it has not been possible to restrict the presence of the acid to the seed storage sites. There has, for example, been leakage, with the new fatty acid being found in cell membranes, where it can be destabilising and can adversely affect their function.

Another approach could be to make efforts to improve agronomic performance of plants like jojoba or evening primrose, but this is patentable and so is not a profitable avenue for the biotech industry to explore.

Producing plastics has been similarly problematic, with adverse effects on growth being common. All GM approaches are dogged by yields that are not economically viable. And the prospect of having industrial oils and plastics in your food as a result of contamination is not appetising.

One notable success in non-food uses has been GM potatoes, with altered starch production. Amylogene, owned by BASF, has produced potatoes high in amylopectin starch, which is more useful to the paper industry than amylose starch.

These GM potatoes are in their final stages of approval in Europe. They are unlikely to be grown in Britain, but could be in eastern Europe, the Nordic countries, Germany, the Netherlands, Belgium and France, all of which grow potatoes for starch production. These are unlikely to pose contamination threats via pollen, but there will need to be systems in place to maintain their separation from the human food chain, as the residue after starch extraction is intended for animal feed.

It is the apparent success of GM cotton, however, that encourages the application of GM to non-food uses. It has attracted little consumer interest and is grown internationally on many millions of hectares. GM insect-resistant cotton has reduced the use of some insecticides in a system which is highly intensive and environmentally damaging. However, the selling of GM cotton as a cure for the ills of pesticide use is eerily familiar to the way in which the pesticides themselves have been sold.

Short-term benefits and high-cost inputs are being promoted by industry salesmen. Loans for seed purchase and second-generation GM crops, if the first-generation cotton fails, are already part of the plan.

Developing countries are the targets for expansion of the GM cotton market in Africa, south-east Asia and South America. So while the prospect of GM food crops being grown in Britain has receded in the short term, the industry has a whole new rationale and a raft of new uses for the technology up its sleeve.

With some of these, such as GM grasses, there will be little comfort that it is happening elsewhere. Inevitable, accidental international movement means they will surely find their way to Britain.

■ Sue Mayer is director of GeneWatch UK, which monitors developments in genetics technologies. More information: www.genewatch.org This article first appeared in the *Guardian*, 5 May 2004.

EU lifts six-year ban on GM corn

By Ian Black in Brussels

The EU is to approve the sale of a brand of genetically modified corn for human consumption – ending a six-year ban that was challenged by the US.

The landmark decision by the European commission will allow the insect and herbicide-resistant Swiss-made product to be sold, even though consumer resistance remains powerful. But growing Syngenta Bt-11 maize in the EU's 25 member states will remain illegal for the moment.

Under new EU rules, canned vegetables have to be clearly labelled as having been harvested from a GM plant, the commission's chief spokesman, Reijo Kemppinen, said 15 May 2004.

Since many supermarket chains require suppliers to guarantee that their products are GM-free, the product is unlikely to be a huge success. The biotech industry trade group, EuropaBio, welcomed the announcement, but conceded that the corn was unlikely to be on sale soon.

David Bowe, Labour's environment spokesman in the European parliament, said: 'This is good news for consumers because it will increase choice and competition. It is also good news because it will increase choice and competition for producers, too.'

But the decision was condemned by Friends of the Earth, the environmental group. 'Scientists cannot agree over [the sweetcorn's] safety and the public does not want it,' a spokesman said. Polls shows some 70% of the European public remains opposed to GM foods.

Critics say Bt-11 has been modified to produce a toxin that is naturally found only in bacteria and that the scientific assessment was undertaken according to outdated rules.

The decision to lift the ban follows the failure of EU governments to agree on the first application submitted under the new labelling rules, which came into effect in April.

The EU has been under pressure from the US and other big exporters to lift the 1998 ban which, they say, is unscientific and illegal under World Trade Organisation rules.

The moratorium came into force when several EU countries said they would reject new GM authorisations until there were stricter laws on testing and labelling.

The commission, backed by Britain, insists that the new rules provide adequate protection for consumers.

Genetically modified foods – the debate moves ahead

Information from EUFIC

The debate over genetically modified (GM) foods has been going on for some years now, with much of the discussion centred on whether or not these foods are safe to eat. Thanks to scientific research, improved understanding of the technology and new regulations, most parties involved in the GM debate now agree that the food and food ingredients derived from currently available genetically modified crops are not likely to present a risk for human health.

Safety assessment

A crucial point to remember when considering the safety of GM foods, is that it is the food as it is consumed that must be examined, not the production process in isolation. This means that the properties and overall safety of the food need to be assessed, just as we do with foods produced using conventional methods. European Union legislation requires that GM products are submitted to a rigorous safety evaluation before authorisation is given for human consumption.

The European Union, the World Health Organisation (WHO) and the Food and Agriculture Organisation (FAO) of the United Nations agreed on a methodology referred to as 'substantial equivalence' as the most practical approach to assess the safety of GM foods and food ingredients.

'Substantial equivalence' focuses on the product rather than the production process. It is a rigorous procedure including a detailed list of parameters and characteristics that need to be considered including molecular characterisation of the genetic modification, agronomic characterisation, nutritional and toxicological assessments.

The 'substantial equivalence' approach acknowledges that the goal of the assessment cannot be to establish absolute safety. The important conclusion is that if, after the evaluation, the safety of the new product is comparable (substantially equivalent) to a conventional counterpart then the level of 'risk' is comparable to foods that we have consumed safely for thousands of years. However, if the GM product has new traits or characteristics that make it no longer substantially equivalent (such as a higher level of a vitamin), then an additional assessment is required. This assessment focuses on the effects the new trait may have on the safety of the new food and may include various types of tests to demonstrate the safety.

Looking ahead

The debate on GM food is far from over. Various environmental issues and the safety assessment of future generations of GM products with unique characteristics are among the issues that will continue to elicit discussion, research and testing.

It is however evident that much progress has been made with regard to developing a consensus on the food safety. As the number of GM products available in the market slowly grows, consumers can be assured that they have been subjected to a rigorous evaluation and that food authorities worldwide agree on their safety for human health.

'Available evidence shows that GM foods are "not likely to present human health risks" and therefore "these foods may be eaten".'

Dr Gro Harlem Brundtland,
WHO Director-General,
28 August 2002.

'Current scientific research confirms the safety of GM food.'

Dr Jacques Diouf,
FAO Director General,
30 August 2002.

References

- FAO/WHO (1991) *Strategies for assessing the safety of foods produced by biotechnology*. Report of a joint FAO/WHO Consultation. WHO. Switzerland.
- OECD (1993) *Safety evaluation of foods produced by modern biotechnology: concepts and principles*. OECD, Paris, France.
- Genetic Modification and Food. *Consumer Health and Safety*. ILSI Europe Concise Monograph Series, 2001.
- WHO, Food Safety Programme, *20 Questions on genetically modified (GM) foods*, 2002.

■ The above information is from EUFIC's website which can be found at www.eufic.org

KEY FACTS

■ Genetic modification involves inserting or changing an organism's genes to produce a desired characteristic. (p. 2)

■ Some see genetically modified crops opening up opportunities in agriculture, food and medicine. For some it's a threat to something basic about ourselves and the natural world – harmful, unnecessary, and benefiting big business at others' expense. (p. 3)

■ There is widespread confusion with just over a third aware that there are GM foods on sale. (p. 5)

■ In the future, GM crops will be able to provide foods with improved nutritional value, longer shelf-life and lower prices. (p. 7)

■ Under EU law, the presence of GM has to be labelled as GM, as long as it can be detected in the final product. The two main GM crops we may be eating are soya and maize (corn). Soya and maize derivatives are found in around 80% of processed foods. (p. 9)

■ Some people do not wish to buy foods produced from animals which have been fed on GM feed. However, the GM food and feed Regulations do not require food labels to indicate if a food has been derived from animals fed on GM feed. (p. 11)

■ Three GM foods or ingredients have been on sale or approved for use in foods in the UK: GM tomatoes, which were sold only as tomato purée; GM soya; GM maize.

■ The Government is entering a consultation process to decide how the UK responds to the controversial subject of GM food. It does so, however, with the public firmly opposed to the introduction of GM foods – more than half the public (56%) opposes GM food, compared to one in seven (14%) who support it. (p. 16)

■ 'The use of GM for non-food crops could bring contamination of food and nature by the back door,' (p. 20)

■ Many foods and other products are marketed as 'natural'. What does this really mean? What is the difference between a 'natural' chemical and any other? (p. 24)

■ Crops can be modified to resist attack by specific pests. This both reduces the amount of pesticide that has to be used and leads to more targeted control, which protects beneficial species such as pollinators and the natural predators of pests. (p. 25)

■ GM industry must pay for damage: According to the government's GM committee, the GM industry will be liable for any damage their GM crops cause. (p. 27)

■ GM crops covered 58 million hectares worldwide in 2002 – an area two and a half times the size of the UK. (p. 28)

■ The four corporations that control most of the GM seed market had a combined turnover from agrochemicals and seeds of $21.6 billion in 2001. (p. 28)

■ Most research and development in GM agriculture is conducted by the private sector. Less than 1% of all GM research is directed at poor farmers. (p. 29)

■ The widespread adoption of GM crops seems likely to exacerbate the underlying causes of food insecurity, leading to more hungry people, not fewer. (p. 30)

■ In total GM crops may have cost the US economy at least $12 billion from 1999 to 2001. The US farm sector has become highly unstable, with record levels of farm bankruptcies. (p. 31)

■ Many non-GM farmers in the USA and Canada are finding it nearly impossible to grow GM-free crops. Seed stocks have become contaminated and there is a high risk of their crop being contaminated by their neighbours', even if they plant normal varieties. (p. 31)

■ Commercialisation of GM crops in the UK would be against consumer demand, would lose Britain its valuable status as a relatively safe haven from GM, and would render organic farming virtually impossible. (p. 32)

■ The FSA funded the world's first and only human feeding trial on GM in 2002. A tiny sample of volunteers eating one GM meal resulted in human gut bacteria taking up GM genes. (p. 33)

■ Some scientists now think they have the answer to what has become agriculture's greatest challenge. On 21 May 2004, a group of world-class researchers in the US announced that their company, FuturaGene, had developed the means to make plants fight harder for their survival in harsh environments. (p. 35)

■ The debate on GM food is far from over. Various environmental issues and the safety assessment of future generations of GM products with unique characteristics are among the issues that will continue to elicit discussion, research and testing. (p. 29)

ADDITIONAL RESOURCES

You might like to contact the following organisations for further information. Due to the increasing cost of postage, many organisations cannot respond to enquiries unless they receive a stamped, addressed envelope.

ActionAid
Hamlyn House
MacDonald Road
London, N19 5PG
Tel: 020 7561 7561
Fax: 020 7272 0899
E-mail: mail@actionaid.org.uk
Website: www.actionaid.org
ActionAid is a unique partnership of people who are fighting for a better world – a world without poverty.

Biotechnology and Biological Sciences Research Council (BBSRC)
Polaris House
North Star Avenue
Swindon, SN2 1UH
Tel: 01793 413200
Fax: 01793 413382
Website: www.bbsrc.ac.uk
Britain's leading funding agency for academic research in the non-medical life sciences.

Consumers' Association
2 Marylebone Road
London, NW1 4DF
Tel: 020 7830 6000
Fax: 020 7830 7600
E-mail: which@which.co.uk
Website: www.which.net
A research and policy organisation providing a vigorous and independent voice for domestic consumers in the UK.

CropGen
PO Box 38589
London, SW1A 1WE
Tel: 0845 602 1793
Fax: 020 7853 2306
E-mail: info@cropgen.org
Website: www.cropgen.org
CropGen's mission is to make the case for GM crops by helping to achieve a greater measure of realism and better balance in the UK public debate about crop biotechnology.

European Food Information Council (EUFIC)
1 Place des Pyramides 75001
Paris, France
Tel: + 33 140 20 44 40
Fax: + 33 140 20 44 41
E-mail: eufic@eufic.org
Website: www.eufic.org
EUFIC was established to provide science-based information on foods and food-related topics.

Food and Drink Federation
6 Catherine Street
London, WC2B 5JJ
Tel: 020 7836 2460
Fax: 020 7836 0580
Websites: www.fdf.org.uk
www.foodfuture.org.uk
www.foodfitness.org.uk
www.foodlink.org.uk
Produces publications and surveys on food and biotechnology.

Food Standards Agency
Aviation House
125 Kingsway
London, WC2B 6NH.
Tel: 020 7276 8000
Websites: www.food.gov.uk
www.foodstandards.gov.uk
www.food.gov.uk/gmdebate
An independent food safety watchdog set up to protect the public's health and consumer interests in relation to food.

GeneWatch UK
The Mill House
Manchester Road
Tideswell, Buxton
Derbyshire, SK17 8LN
Tel: 01298 871898
Fax: 01298 871898
Website: www.genewatch.org
Monitors developments in genetic technologies from a public interest, environmental protection and animal welfare perspective.

John Innes Centre
Norwich Research Park
Colney, Norwich, NR4 7UH
Tel: 01603 450000
Fax: 01603 450045
Websites: www.jic.bbsrc.ac.uk
www.gmissues.org
Europe's premier independent centre for research and training in plant and microbial science.

Society, Religion and Technology Project
Church of Scotland,
John Knox House, 45 High Street
Edinburgh, EH1 1SR
Tel: 0131 556 2953
Fax: 0131 556 7478
E-mail: srtp@srtp.org.uk
Website: www.srtp.org.uk
A project of the Church of Scotland to examine ethical issues emerging from modern technology and engaging with key scientists and policy makers.

The Soil Association
Bristol House
40-56 Victoria Street
Bristol, BS1 6BY
Tel: 0117 929 0661
Fax: 0117 925 2504
E-mail: info@soilassociation.org
Website: www.soilassociation.org
Works to educate the general public about organic agriculture, gardening and food, and their benefits for both human health and the environment.

Totnes Genetix Group
PO Box 77
Totnes, Devon, TQ9 5UA
Tel: 01803 840098
Fax: 01803 864591
E-mail: info@togg.org.uk
Website: www.togg.org.uk
Togg's aim is to inspire and empower ourselves and others to create a balanced, diverse, sustainable and ecologically sound food production and distribution system based on the principles of organic farming.

World Health Organization (WHO)
20 Avenue Appia
1211-Geneva 27, Switzerland
Tel: + 41 22 791 2111
Fax: + 41 22 791 3111
E-mail: info@who.ch
Website: www.who.int
WHO works to make a difference in the lives of the world's people by enhancing both life expectancy and health expectancy.

INDEX

ACKNOWLEDGEMENTS

The publisher is grateful for permission to reproduce the following material.

While every care has been taken to trace and acknowledge copyright, the publisher tenders its apology for any accidental infringement or where copyright has proved untraceable. The publisher would be pleased to come to a suitable arrangement in any such case with the rightful owner.

Overview

GM *basics*, © Crown copyright is reproduced with the permission of Her Majesty's Stationery Office, *DNA*, © Crown copyright is reproduced with the permission of Her Majesty's Stationery Office, *Plant cell*, © Crown copyright is reproduced with the permission of Her Majesty's Stationery Office, *Genetically modified food*, © Society, Religion and Techology Project, GM *crops*, © GeneWatch UK, GM *dilemmas*, © Consumers' Association, *Consumers and GM food*, © Consumers' Association, *Benefits*, © CropGen, *Issues of concern*, © World Health Organization, *Genetically modified organisms*, © Foodaware, *Regulations*, © Food and Drink Federation.

Chapter One: GM Food

GM *food – opening up the debate*, © Food Standards Agency, *Labelling rules in the UK*, © Food Standards Agency, *Why it's time for GM Britain*, © Guardian Newspapers Limited 2004, *Continuing opposition to GM foods*, © MORI, *Public attitudes to GM food*, © MORI, *Is GM food safe?*, © Crown copyright is reproduced with the permission of Her Majesty's Stationery Office, *What is being done about the potential risks?*, © Biotechnology and Biological Sciences Research Council (BBSRC), *Contamination threats*, © GeneWatch UK, GM *products 'slip into foods'*, © 2004 Associated New Media, *Genetic modification of plants and food crops*, © 2000-2004, John Innes Centre, Norwich, UK, *What are the potential benefits?*, © Biotechnology and Biological Sciences Research Council (BBSRC), *Myths & GM*, © Monsanto, GM – *update and suggestions for action*, © Soil Association 2000-2004, GM *crops – going against the grain*, © ActionAid, *Area and type of GM crops grown in developing countries, 2002*, © James C, ISAAA, GM *losers*, © Totnes Genetics Group, *Closing the stable door after the GM horse has bolted*, © Soil Association 2000-2004, *Monsanto abandons GM wheat project*, © Guardian Newspapers Limited 2004, *Breakthrough may bring life to barren earth*, © Guardian Newspapers Limited 2004, *The gene revolution*, © CropGen, *Against the grain*, © Sue Mayer, *EU lifts six-year ban on GM corn*, © Guardian Newspapers Limited 2004, *Genetically modified foods – the debate moves ahead*, © EUFIC.

Photographs and illustrations:

Pages 1, 21, 31, 37: Simon Kneebone; pages 9, 25, 33: Don Hatcher; pages 10, 17: Pumpkin House; pages 12, 27: Bev Aisbett; pages 14, 24, 36: Angelo Madrid.

Craig Donnellan
Cambridge
September, 2004